COLLINS POCKET REFERENCE

READY REFERENCE

The Diagram Group

D0775766

HarperCollins*Publishers*

HarperCollins Publishers
P.O. Box, Glasgow G4 0NB

A Diagram book first created by Diagram Visual Information
Limited of 195 Kentish Town Road, London NW5 8SY

First published 1994

Reprint 10 9 8 7 6 5 4 3 2 1 0

© Diagram Visual Information Limited 1994

ISBN 0 00 470322 7

A catalogue record for this book is available
from the British Library

Printed in China

Foreword

How heavy is a gold atom? What does the medical abbreviation 'ECG' mean? What currency is used in Uzbekistan? All these questions and many more are answered in the *Collins Pocket Reference Ready Reference*. Subjects as diverse as geometry and geography, the Russian alphabet and cooking measures are included.

This volume is divided into 14 sections, each of which provides essential information on the main units of measurement or features of a particular topic. When relevant, individual chapters include conversion formulas, e.g. metric and imperial equivalents, plus conversion tables that give immediate visual reference.

The *Collins Pocket Reference Ready Reference* is an indispensable guide to the international variety in units of measurement, symbols, common and specific abbreviations, and astronomical and geographical data. It is an invaluable and essential companion, whether for school, the office or home.

How to use this book

The *Collins Pocket Reference Ready Reference* is divided into 14 sections, each of which is devoted to a particular category of facts and figures. If you know which category you wish to explore, merely turn to the table of contents to find the relevant page number.

Unit conversion index

In this book, there are tables for converting units from the imperial system of measurement to the metric system (and vice versa), and for converting one type of unit to another within the same system. The Unit Conversion Index (page 8) enables you to refer quickly to the tables in which a particular unit is converted.

Formulas

Within sections 2 to 6 and 8, you will find a selection of conversion formulas. These are easy-to-use formulas for common conversions; you will need to use a calculator for most of them, although many are simple, approximate conversions.

Conversion tables

Each group of units has its conversion tables: pages of quick-reference tables for all imperial and metric measurements from metres to feet, grains to grams. These are particularly handy if you do not have a calculator. It would be impossible to accommodate tables listing every possible conversion, so the material included is not exhaustive.

Note also that the figures in the conversion tables are rounded up or down to the third decimal place, and so are not always exact.

Contents

CONTENTS

13: SYMBOLS, CODES AND ALPHABETS

14: ABBREVIATIONS

Unit conversion index

Glossary

Scientific abbreviations and abbreviations used in measuring are shown on pages 310 – 312.

acre A measure of land; originally the amount of land that a yoke of oxen could plough in a day. Equal to 4840 yd^2.

amu *see* Atomic mass unit.

ampere (A) The unit for measuring electric current.

ångström (Å) A unit of length, used mainly to measure the wavelength of light. Named after the Swedish physicist A.J. Ångström (1814–74). Equal to 10^{-10} m (10^{-8} cm).

apothecaries' system A system of weights used especially by pharmacists.

are (a) A unit of measure equal to an area of 10×10 m (1 a = 100 m^2). *See also* Hectare (ha): 100 a = 1 ha.

astronomical unit (au or AU) A unit of measure based on the distance between the Earth and the Sun. Approximately equal to 1.5×10^8 km.

atomic mass unit (amu)

    chemical A unit of mass equal to $^1/_{16}$ of the weighted mass of the three naturally occurring neutral oxygen isotopes.
    1 amu chemical = $(1.660 \pm 0.000\ 05) \times 10^{-27}$ kg. Formerly called the atomic weight unit.

    international A unit of mass equal to $^1/_{12}$ of the mass of a neutral carbon-12 atom. 1 amu international = $(1.660\ 33 \pm 0.000\ 05) \times 10^{-27}$ kg.

physical A unit of mass equal to $\frac{1}{16}$ of the mass of an oxygen atom. 1 amu physical = 1.660×10^{-27} kg.

atto- In the UK, a prefix meaning a trillionth (10^{-18}); in the US, meaning a quintillionth (10^{-18}). For example, in the UK 1 attometre = 1 trillionth of a metre; in the US, 1 attometre = 1 quintillionth of a metre.

avoirdupois system A system of weights based on the 16-ounce pound and the 16-dram ounce.

baker's dozen A counting unit equal to 13.

barleycorn A unit of measure of length equal to $\frac{1}{3}$ in.

billion (bil) In the UK, equal to 10^{12}; in US, equal to 10^9. Commonly now also used in the UK to mean 10^9.

bolt A measure of length, usually for fabric. In the UK, a bolt of cloth equals 42 yd; in the US, a bolt of wallpaper equals 16 yd and a bolt of cloth equals 40 yd.

British thermal unit (Btu) Measure of heat needed to raise the temperature of one pound of water by 1 °F. Equal to 252 calories.

bushel (bu) A measure of dry volume. In the UK, 1 bu = 8 gal (64 UK pt); in the US, 1 bu = 8 gal (64 US pt). The measures are not to be confused: 1 UK bu = 1.03 US bu.

calibre A unit of length used to measure the diameter of a tube or the bore of a firearm, in $\frac{1}{100}$ in or $\frac{1}{1000}$ in increments.

calorie (cal) A measure of heat energy representing the amount of heat needed to raise 1 g of water by 1 °C. Also called 'small calorie': 1000 cal = 1 kcal or Cal. *See also* Joule; Kilocalorie.

carat A unit of weight equal to 200 mg (3.1 grains). Also used as a measure of gold purity (per 24 parts gold alloy).

centi- Prefix meaning 100 or $\frac{1}{100}$; e.g., a centilitre (cl) is a unit of volume equal to $\frac{1}{100}$ (0.01) litre.

centrad A measure of a plane angle, especially used to measure the angular deviation of light through a prism. 1 centrad = $\frac{1}{100}$ (0.01) radian.

century A measure of time equal to 100 years.

chain A measure of length equal to 22 yd. Also known as Gunter's chain.

   engineer's chain A measure of length equal to 100 ft.

   nautical chain A measure of length equal to 15 ft.

   square chain A measure of area equal to 484 yd^2.

chaldron A measure of volume. In the UK, 1 chaldron = 36 UK bu (288 gal); in the US, 1 chaldron = 36 US bu.

cord A unit of dry volume, especially used for timber. Equal to 128 ft^3.

cubic units (cu or ^{3}**)** These are arrived at by multiplying a number by itself twice. With a three dimensional object, the height, breadth and length are multiplied togther to give its volume, which is measured in cubic units.

cubit A unit of length approximately equal to 18 in. Originally based on the distance from the tip of the middle finger to the elbow.

cup A measure of volume (either liquid or solid) used especially in cooking. In the UK, 1 cup = ½ UK pt (16 tbsp); in the US, 1 cup = ½ US pt (16 tbsp). The two should not be confused: 1 UK cup = 1⅕ US cups.

day

    mean solar day A measure of time representing the interval between consecutive passages of the Sun across the meridian, averaged over 1 year.
    1 day = 24 hr (86 400 s).

    sidereal day A measure of time approximately equal to 23 hr, 56 min, 4.09 s. A sidereal day represents the time needed for one complete rotation of the Earth on its axis.

deca- Prefix meaning ten; e.g., a decametre is a measure of length equal to 10 m.

decade A measure of time equal to 10 years.

deci- Prefix meaning $\frac{1}{10}$; e.g., a decilitre (dl) is a measure of liquid volume equal to $\frac{1}{10}(0.01)$ litre.

decibel (dB) A measure of relative sound intensity.

degree (°)

    geometrical A unit of measure of plane angle equal to $\frac{1}{360}$ of the circumference of a circle (1 circle = 360°).

    temperature A measure of temperature difference representing a single division on the temperature scale. The centigrade scale has 100 equal degrees; the Fahrenheit scale has 212 equal degrees.

digit One of ten Arabic symbols representing numbers 0 to 9. Also used in astronomy as a unit of measure equal to $\frac{1}{12}$ the diameter of the Sun or Moon. Used in ancient Egypt as a measure of length: 1 digit = 1 finger width.

douzieme A unit of length equal to $\frac{1}{12}$ line.

dozen A counting unit equal to 12.

drachm A unit of mass in the apothecaries' system. 1 drachm = $\frac{1}{8}$ apothecaries' ounce (60 grains).

dram (dr) A unit of mass equal to $\frac{1}{16}$ oz.

fluid dram A unit of liquid volume. In the UK,
1 dr = ⅛ UK fl oz; in the US, 1 dr = ⅛ US fl oz.
The two should not be confused:
1 UK fl dr = 0.960 759 US fl dr.

dry Used in US to distinguish measures of dry (solid)
volume as opposed to liquid (fluid) volume. For
example, in the UK, the pint measures both dry and
liquid volume. In the US, 1 fl pt = ⅛ US gal; 1 dry pt =
¹⁄₆₄ US bu. 1 US dry pt ≈ 0.969 UK pt ≈ 1.163 US fl pt.

dyne A unit of force equal to that needed to produce
acceleration of 1 cm per second in a mass of 1 g.
Replaced by the newton (N): 1 dyne = 10^{-5} N.

electronvolt (eV) A unit of energy measurement
representing the energy acquired by an electron in
passing through a potential difference of 1 volt.
1 eV = $(1.6 \pm 0.000\ 07) \times 10^{-19}$ J.

erg A unit of energy measurement equal to the energy
produced by a force of 1 dyne through a distance of
1 cm. Replaced by the joule, 1 erg = 10^{-7} J.

fathom (fm) Unit of length, especially used to measure
marine depth. 1 fm = 6 ft. Originally based on the span
of two outstretched arms.

feet per minute A unit of velocity representing the
number of feet travelled in 1 min.

femto- In the UK, a prefix meaning 1 thousand
billionth (10^{-15}); in the US, meaning 1 quadrillionth
(10^{-15}).

firkin A unit of volume, used especially to measure
beer or ale. In the UK, 1 firkin = 9 UK gal; in the US,
1 firkin = 9.8 US gal.

fluid Used to distinguish units of liquid (fluid) volume as opposed to dry (solid) volume.

fluid dram *see* Dram.

fluid ounce *see* Ounce.

foot (ft) A unit of length equal to 12 in.

furlong (fur) Unit of length equal to ⅛ mi (660 ft).

gallon (gal) A unit of liquid volume. In the UK, 1 gal = 8 UK pt; in the US, 1 gal = 8 US pt. The two should not be confused: 1 UK gal = 1.2 US gal.

> **Winchester wine gallon (WWG)** A unit of volume used for wine, honey, or other liquids. Equal to 0.83 UK gal.

gauge A unit of length used to measure the diameter of a shotgun bore; e.g., 6-gauge equals 23.34 mm. Originally based on the number of balls, of certain size, contained in 1 lb of shot.

giga- In the UK, a prefix meaning 1 thousand million (10^9); in the US, meaning 1 billion (10^9). For example, in the UK, 1 gigametre = 1 thousand million metres; in the US, 1 gigametre = 1 billion metres.

gill A unit of liquid volume. In the UK, 1 gill = ¼ UK pt; in US (gi), 1 gi = ¼ US fl pt. The two should not be confused: 1 UK gill = ½ US gi.

grade (^{g}) A measure of plane angle in geometry. $1^g = 0.9°$.

grain (gr) A unit of mass measurement, used especially in the apothecaries' system. 1 grain = 1/7000 lb (avoirdupois); 480 grains = 1 ounce troy; 24 grains = 1 pennyweight.

gram (g) A unit of mass or volume measurement. 1 g = 0.001 kg.

gross A counting measure equal to 144 (or 12 dozen).

hand A unit of length, used especially to measure
horses' height. 1 hand = 4 in.

hectare (ha) A measure of area, usually of land, equal
to 10 000 m^2.

hecto- Prefix meaning 100; e.g., a hectometre (hm) is a
unit of length equal to 100 m.

hertz (Hz) A unit of frequency measurement equal to
1 cycle per second.

horsepower (hp) A unit of work representing the
power needed to raise 550 lb by 1 ft in 1 s.

   metric horsepower A unit of power representing
   that needed to raise a 75-kg mass 1 m in 1 s.

hour (hr) A unit of time measurement equal to 60 min
(3600 s).

hundredweight (cwt) A unit of mass.
   1 hundredweight =112 lb; 1 hundredweight troy =
   100 pounds troy; 1 hundredweight = 4 quarters.
   short hundredweight (sh cwt) US name for the
   UK quintal or cental.
   1 short hundredweight = 100 lb.

inch (in) A unit of length equal to $\frac{1}{12}$ ft.

inches per second A unit of velocity representing the
number of inches travelled in 1 s.

joule (J) A unit of energy equal to the work done when
a force of 1 newton is moved through a distance of 1
m. Used instead of calorie: 1 J = 0.239 cal. Named
after J.P. Joule (1818–89).

keg A unit of volume, used especially for beer, approximately equal to 30 gal. Also used as a measure of weight for nails, equal to 100 lb.

kelvin (K) A scale of temperature measurement in which each degree is equal to $\frac{1}{273.16}$ of the interval between 0 K (absolute zero) and the triple point of water. K = °C + 273.16. Named after William Thomson, Lord Kelvin (1824–1907).

kilo- Prefix meaning 1000; e.g., a kilogram (kg) is a unit of volume measurement equal to 1000 g.

kilocalorie (kcal or Cal) A unit of energy measurement representing the amount of heat required to raise 1 kg of water by 1 °C. Also called the 'international calorie'. 1 kcal = 1000 cal. *See also* Calorie.

kilogram *see* Kilo-

kilometre (km) A unit of length equal to 1000 m.

kiloparsec A unit of distance used to measure distance between galactic bodies. 1 kiloparsec = 3260 light years (ly).

kilowatt (kW) A unit of power equal to 1000 watts (W).

kilowatt-hour (kWh) A unit of energy equal to the energy expended when a power of 1 kW is used for 1 hr.

knot (kn) A nautical unit of speed measurement equal to the velocity at which 1 n mi is travelled in 1 hr. 1 kn = 6076 ft per hour.

lakh An Indian counting unit equal to 100 000.

lambda (λ) A unit of volume measurement. 1 λ = 1 microlitre (10^{-6} litre).

league A unit of length equal to 3 mi.

light year (ly) A unit of length (distance) representing the distance travelled by electromagnetic waves (light) through space in 1 year. 1 light year = 9.4605×10^{12} km (or, in the UK, 9 billion miles; in the US, 9 trillion miles).

line A unit of length equal to $\frac{1}{12}$ in; 4 lines = 1 barley-corn. It can also be used to measure button diameters, when 1 line = $\frac{1}{40}$ in.

litre (l) A unit of volume measurement equal to the volume of 1 kg of water at its maximum density. 1 litre = 1000 cm^3.

magnum A measure of volume, used especially for wine or champagne. In the UK, 1 magnum = $\frac{2}{5}$ UK gal; in the US, 1 magnum = $\frac{2}{5}$ US gal.

mega- Prefix meaning 1 million; e.g., a megaton is a unit of weight equal to 1 million tons.

megahertz (MHz) A unit of frequency (for radio) equal to 1 million cycles per second.

metre (m) A unit of length equal to 100 cm.

metres per minute (m/min) A unit of velocity measurement representing the number of metres travelled in 1 min.

metric system A system of measurement based on the metre.

micro- Prefix meaning 1 millionth; e.g., a microlitre is a unit of volume equal to 1 millionth of a litre.

micron (μm) A unit of length equal to $\frac{1}{1000}$ (0.001) mm. Also called the micrometre.

mile (mi) A unit of length equal to 1760 yd. Also called the statute mile in the UK.

nautical mile (n mi) A unit of length used in navigation. In the UK, 1 n mi = 6080 ft; in the metric system, 1 n mi (international) = 1852 m. Also called the geographical mile.

sea mile A unit of length distinguished from the nautical mile. 1 sea mile = 1000 fathoms (6000 ft).

miles per hour (mph) A unit of velocity representing the number of miles travelled in 1 hr.

millennium A period of time equal to 1000 years.

milli- Prefix meaning 1 thousandth or $\frac{1}{1000}$; e.g., 1 millimetre (mm) is a unit of length equal to $\frac{1}{1000}$ (0.001) m.

minim A unit of volume, usually for liquids. In the UK, 1 minim = $\frac{1}{480}$ UK fl oz; in the US, 1 minim = $\frac{1}{480}$ US fl oz. The two should not be confused: 1 UK minim = 0.961 US minim.

minute

    geometric ($'$) A unit of measure for plane angles. $1' = \frac{1}{60}°$.

    time (m or min) A unit of time measurement equal to 60 s. 60 min = 1 hr.

month

    lunar A unit of time equal to 4 weeks (2 419 200 s).

    sidereal *see* Year, sidereal.

    tropical *see* Year, tropical.

nano- In the UK, a prefix meaning 1 thousand millionth (10^{-9}); in the US, meaning 1 billionth (10^{-9}). For example, in the UK, 1 nanometre = 1 thousand millionth of a metre; in the US, 1 nanometre = 1 billionth of a metre.

nautical mile *see* Mile.

newton (N) A unit of force equal to that creating an acceleration of 1 m per second when applied to a mass of 1 kg. This unit has replaced the dyne: $1 N = 10^5$ dynes. Named after Isaac Newton (1642–1727).

ohm (Ω) A unit of electrical resistance. One ohm equals the resistance across which a potential difference of 1 volt produces a current flow of 1 ampere. Named after G.S. Ohm (1787–1854).

ounce (oz) A unit of mass equal to $\frac{1}{16}$ lb.

    fluid ounce A unit of liquid volume measurement. In the UK, 1 fl oz = $\frac{1}{20}$ UK pt; in the US, 1 fl oz = $\frac{1}{16}$ US pt.

    metric ounce A unit of mass equal to 25 g. Also called a Mounce.

    ounce troy A unit of mass in the troy system. Equal to $\frac{1}{12}$ pound troy.

pace A unit of length/distance equal to 5 ft, used in ancient Rome.

palm A unit of length used in ancient Egypt, equal to the width of an average palm of the hand (4 digits).

parsec (pc) A unit of length used for measuring astronomical distances. 1 parsec = 3.26 light years (ly).

pascal (pa) A unit of pressure equal to the force of 1 N acting over an area of 1 m^2.

peck (pk) A unit of dry volume. In the UK, 1 peck = 2 UK gal; in the US, 1 peck = 2 US gal. The two should not be confused: 1 UK peck ≈ 1.032 US peck.

pennyweight (dwt) A unit of weight in the troy system equal to $\frac{1}{20}$ ounce troy (25 grains).

perch A unit of length equal to 5½ yd. Also called a pole or a rod.

pi (π) Symbol and name representing the ratio of a circle's circumference to its diameter. Its value is approximately 3.14.

pica A unit of length, used by printers, approximately equal to ⅙ in.

pico- In the UK, a prefix meaning 1 billionth (10^{-12}); in the US, meaning 1 trillionth (10^{-12}). For example, in the UK, 1 picometre = 1 billionth of a metre; in the US, 1 picometre = 1 trillionth of a metre.

pint (pt) A unit of volume. In the UK, a pint measures either dry or liquid volume: 1 pt = ⅛ UK gal. In the US, two kinds of pint are used: 1 fl pt = ⅛ US gal; 1 dry pt = ¹⁄₆₄ US bu. These two should not be confused: 1 UK pt ≈ 1.2 US fl pt ≈ 1.03 US dry pt.

point A unit of length, used especially by printers, approximately equal to ¹⁄₇₂ in.

pole Unit of length equal to 5½ yd. *See also* Perch; Rod.

pound (lb) A unit of mass equal to 453.59 g.

> **force pound** A unit of force equal to 32.174 poundals. Also called pound-force.

> **pound troy (lb tr)** A unit of mass in the troy system. 1 pound troy = 12 ounces troy.

poundal A unit of force equal to that needed to give an acceleration of 1 ft per second to a mass of 1 lb.

PSI Pounds per square inch: a unit for measuring pressure. 1 PSI equals the pressure resulting from a force of 1 force pound acting over an area of 1 in². *See also* Pound.

quart (qt) A unit of volume, usually for liquids. In the UK, 1 qt = 2 UK pt; in the US, 1 qt = 2 US fl pt. The two should not be confused: 1 UK qt ≈ 1.2 US qt.

> **dry quart (dry qt)** A unit of measure for dry (solid) volume in US.

> **reputed quart** A unit of volume, used especially for wine, equal to ⅙ of a Winchester wine gallon.

> **Winchester quart** A unit of fluid volume equal to 2.5 litres.

quarter (qr)

> **mass quarter** A unit of mass. In the UK, 1 quarter = ¼ UK hundredweight (28 lb); in the US, 1 quarter = ¼ US ton (500 lb).

> **quarter troy (qr tr)** A unit of weight equal to 25 troy pounds.

> **volume quarter** A unit of volume, in the UK, equal to 8 UK bu.

quintal (q) A unit of mass equal to 100 kg or 100 lb. Called the short hundredweight in the US.

rad A short form of radian, a unit of measure for plane angles. *See also* Centrad.

ream A unit of volume, used to measure paper in bulk. 1 ream equals about 500 sheets.

rod

> **area rod** A unit of area equal to 30¼ yd². Also called a square perch or a square pole.

> **length rod** A unit of length equal to 5½ yd. *See also* Perch; Pole.

rood A unit of area equal to ¼ acre (1210 yd²).

score A counting unit equal to 20.

scruple A unit of mass in apothecaries' system equal to 20 grains.

second A unit of time equal to $\frac{1}{60}$ minute.

    geometric (′) A measure of plane angle equal to $\frac{1}{360}°$ and $\frac{1}{60}″$.

    orbital A unit of time equal to $\frac{1}{31\,557}$ of the tropical year 1900. Also called Ephemeris second.

    sidereal A unit of time equal to $\frac{1}{86\,400}$ of the interval needed for one complete rotation of the Earth on its axis.

square units (sq or ²) These are arrived at by multiplying a number by itself once. To find the area of, e.g. a square or rectangle, length and breadth are multiplied together to give the area, which is measured in square units.

stere A unit of volume, especially used for measuring timber. 1 stere = 1 m³.

stone (st) A unit of mass used in the UK. 1 st = 14 lb.

tablespoon (tbsp) A unit of volume used in cooking and equal to 1.5 centilitres (3 tsp). 16 tbsp = 1 cup.

teaspoon (tsp) A unit of volume used in cooking and equal to 0.5 centilitre. 3 tsp = 1 tbsp.

tera- In the UK, a prefix meaning 1 billion (10^{12}); in the US, meaning 1 trillion (10^{12}). For example, in the UK, 1 terametre = 1 billion metres; in the US, 1 terametre = 1 trillion metres.

ton A unit of mass. In the UK, 1 ton = 2240 lb. Called a long ton in the US. In the US, 1 ton = 2000 lb. Called a short ton in the UK.

ton troy (ton tr) A unit of mass equal to 2000 pounds troy.

tonne (t) A unit of mass equal to 1000 kg. Also called a metric ton.

tonne of coal equivalent A measure of energy production/consumption based on the premise that 1 tonne of coal provides 8000 kilowatt-hours (kWh) of energy.

trillion In UK, equal to 10^{18}; in US, equal to 10^{12}.

troy system A system of mass measurement based on the 20-ounce pound and the 20-pennyweight ounce.

volt (V) A unit of electromotive force and potential difference. Equal to the difference in potential between two points of a conducting wire carrying a constant current of 1 ampere (A), when the power released between the points is 1 watt (W). Named after Alessandro Volta (1745–1827).

watt (W) A unit of power equal to that available when 1 J of energy is expended in 1 s.
1 W = 1 volt-ampere; 746 W = 1 horsepower (hp). Named after James Watt (1736–1819).

X-unit (x or XU) A unit of length used especially for measuring wavelength. 1 x-unit $\approx 10^{-3}$ ångström (10^{-13} m).

yard (yd) A unit of length equal to 3 ft (36 in).

yards per minute (ypm) A unit of velocity representing the number of yards travelled in 1 min.

year A unit of time measurement determined by the revolution of the Earth around the Sun.

anomalistic year Equals the time interval between two consecutive passages of the Earth through its perihelion (365 days, 6 hr, 13 min, 53 s).

sidereal year Equals the time in which it takes the Earth to revolve around the Sun from one fixed point (usually a star) back to the same point (365 days, 6 hr, 9 min, 9 s).

tropical year Equals the time interval between two consecutive passages of the Sun, in one direction, through the Earth's equatorial plane (or from vernal equinox to vernal equinox; 365 days, 5 hr, 48 min, 46 s).

Unit systems

International System of Units
The International System of Units (or Système
International d'Unités – SI) is the current form of the
metric system that has been in use since 1960. In the
UK, the SI system is used in education, science and,
increasingly, in everyday life.

Base units
There are seven base units in SI:

Unit	Symbol	Quantity
metre	m	length/distance
kilogram	kg	mass
ampere	A	electric current
kelvin	K	temperature
candela	cd	luminosity
second	s/sec	time
mole	mol	amount of substance

Supplementary units
There are also two supplementary units:

Unit	Symbol	Quantity
radian	rad	plane angle
steradian	sr	solid angle

Derived units
In addition, the system uses derived units, which are
expressed in terms of the seven base units above. For
example, velocity is given in metres per second (m/s,
ms^{-1}). Other derived units in SI are referred to by
special names. For example, the watt (W) is a unit of
power; the joule (J) is a unit of energy.

Imperial system of units

Length 1 inch (in or ")			
1 foot (ft or ')	12 in		
1 yard (yd)	36 in	3 ft	
1 fathom (fm)	72 in	6 ft	2 yd
1 chain (ch)		66 ft	22 yd
1 furlong			220 yd
1 mile (mi)		5 280 ft	1760 yd
Area 1 square inch (in²)			
1 square foot (ft²)	144 in²		
1 square yard (yd²)		9 ft²	
1 square chain (ch²)			484 yd²
1 acre			4840 yd²
1 square mile (mi²)	640 acres		
Volume 1 cubic in (in³)			
1 fluid ounce (fl oz)			
1 pint (pt)	20 fl oz		
1 quart (qt)	40 fl oz	2 pt	
1 gallon (gal)		8 pt	4 qt
1 cubic foot (ft³)	1728 in³		
1 cubic yard (yd³)	46 656 in³	27ft³	
Weight 1 grain (gr)			
1 ounce (oz)	437.5 gr		
1 pound (lb)		16 oz	
1 stone (st)			14 lb
1 ton		35 840 oz	2240 lb

1: Numbers

Named numbers

Many numbers have names. Some of these names are in everyday use, others apply in more specialised areas such as music and multiple births; and for sums of money. Some names for specialised numbers have the same first part (prefix). These prefixes indicate the number to which the name refers.

Everyday use	
$^1/_{10}$	Tithe
2	Pair, couple, brace
6	Half a dozen
12	Dozen
13	Baker's dozen
20	Score
50	Half century
100	Century
144	Gross

Musicians	
1	Soloist
2	Duet
3	Trio
4	Quartet
5	Quintet
6	Sextet
7	Septet
8	Octet

Multiple births	
2	Twins
3	Triplets
4	Quadruplets (quads)
5	Quintuplets (quins)
6	Sextuplets

Slang for money	
£25	Pony
£100	Century
£500	Monkey
£1000	Grand

Numerical prefixes

Prefixes in numerical order

$1/10$ Deci-

$1/2$ Semi-, hemi-, demi-

1 Uni-

2 Bi-, di-

3 Tri-, ter-

4 Tetra-, tetr-, tessera-, quadri-, quadr-

5 Pent-, penta-, quinqu-, quinque-, quint-

6 Sex-, sexi-, hex-, hexa-

7 Hept-, hepta-, sept-, septi-, septem-

8 Oct-, octa-, octo-

9 Non-, nona-, ennea-

10 Dec-, deca-

11 Hendeca-, undec-, undeca-

12 Dodeca-

15 Quindeca-

20 Icos-, icosa-, icosi-

Prefixes in alphabetical order

Bi-,	**2**	Pent-, penta-	**5**
Dec-, deca-	**10**	Quadr-, quadri-	**4**
Deci-	$1/10$	Quindeca-	**15**
Demi-	$1/2$	Quinqu-, quinque-	**5**
Di-	**2**	Quint-	**5**
Dodeca-	**12**	Semi-	$1/2$
Ennea-	**9**	Sept-, septem-, septi-	**7**
Hemi-	$1/2$	Sex-, sexi-	**6**
Hendeca-	**11**	Ter-	**3**
Hept-, Hepta-	**7**	Tessera-	**4**
Hex-, hexa-	**6**	Tetr-, tetra-	**4**
Icos-, icosa-, icosi-	**20**	Tri-	**3**
Non-, nona-	**9**	Undec-, undeca-	**11**
Oct-, octa-, octo-	**8**	Uni-	**1**

Prefixes and their values

Prefixes in order of value	Value
*Atto-	0.000 000 000 000 000 001
*Femto-	0.000 000 000 000 001
*Pico-	0.000 000 000 001
*Nano-	0.000 000 001
*Micro-	0.000 001
*Milli-	0.001
*Centi-	0.01
*Deci-	0.1
Semi-, hemi-, demi-	0.5
Uni-	1
Bi-, di-	2
Tri-, ter-	3
Tetra-, tetr-, tessera-, quadri-, quadr-	4
Pent-, penta-, quinqu-, quinque-, quint-	5
Sex-, sexi-, hex-, hexa-	6

* approved for use with the SI system

Prefixes in numerical order	Value
Hept-, hepta-, sept-, septi-, septem-	7
Oct-, octa-, octo-	8
Non-, nona-, ennea-	9
Dec-, deca-	10
Hendeca-, undec-, undeca-	11
Dodeca-	12
Quindeca-	15
Icos-, icosa-, icosi-	20
Hect-, hecto-	100
*Kilo-	1000
Myria-	10 000
*Mega-	1 000 000
*Giga-	1 000 000 000
*Tera-	1 000 000 000 000

Historic number systems

Different civilizations have developed their own systems for writing numbers. Here we show numerals from eight such systems.

	Roman	Arabic	Chinese	Hindu
1	I	١	一	?
2	II	٢	二	?
3	III	٣	三	?
4	IV	٤	四	?
5	V	٥	五	?
6	VI	٦	六	?
7	VII	٧	七	?
8	VIII	٨	八	?
9	IX	٩	九	?
10	X	١٠	十	?o
50	L	٥٠	五十	?o
100	C	١٠٠	百	?oo
500	D	٥٠٠	百五	?oo
1000	M	١٠٠٠	千	?ooo

Babylonian	Egyptian	Hebrew	Japanese
𒁹	I	א	一
𒈫	II	ב	二
𒐈	III	ג	三
𒐉	IIII	ד	四
𒐊	III / II	ה	五
𒐋	III / III	ו	六
𒐌	IIII / III	ז	七
𒐍	IIII / IIII	ח	八
𒐎	III / III / III	ט	九
𒌋	∩	י	十
𒐏	∩∩∩ / ∩∩	כ	五十
𒐐	𓂭	ק	百
𒐏𒁹	𓆼𓆼𓆼 / 𓆼𓆼	תק	五百
𒁹 𒌋	𓆼	אלף	千

Roman number system

The Roman numeral system is a method of notation in which the capitals are modelled on ancient Roman inscriptions. The numerals are represented by seven capital letters of the alphabet:

I	one
V	five
X	ten
L	fifty
C	one hundred
D	five hundred
M	one thousand

These letters are the foundation of the system; they are combined in order to form all numbers. If a letter is preceded by another of lesser value (e.g., IX), the value of the combined form is the difference between the values of each letter (e.g., IX = X (10) – I (1) = 9).

To determine the value of a string of Roman numbers (letters), find the pairs in the string (those beginning with a lower value) and determine their values, then add these to the values of the other letters in the string:

MCMXCI = M+CM+XC+I = 1000+900+90+1 = 1991

A dash over a letter multiplies the value by 1000 (e.g. $\overline{V}$ = 5000).

1 I	12 XII	35 XXXV	100 C
2 II	13 XIII	40 XL	200 CC
3 III	14 XIV	45 XLV	300 CCC
4 IV or IIII	15 XV	50 L	400 CD
5 V	16 XVI	55 LV	500 D
6 VI	17 XVII	60 LX	600 DC
7 VII	18 XVIII	65 LXV	700 DCC
8 VIII	19 XIX	70 LXX	800 DCCC
9 IX	20 XX	75 LXXV	900 CM
10 X	25 XXV	80 LXXX	1000 M
11 XI	30 XXX	90 XC	2000 MM

Mathematical symbols

+	plus or positive	$\geq$	greater than or equal to
−	minus or negative	$\leq$	less than or equal to
±	plus or minus, positive or negative	$\gg$	much greater than
×	multiplied by	$\ll$	much less than
÷	divided by	$\sqrt{}$	square root
=	equal to	∞	infinity
≡	identically equal to	$\propto$	proportional to
≠	not equal to	$\sum$	sum of
≢	not identically equal to	$\prod$	product of
≈	approximately equal to	Δ	difference
∼	of the order of or similar to	$\therefore$	therefore
>	greater than	$\angle$	angle
<	less than	$\parallel$	parallel to
≯	not greater than	$\perp$	perpendicular to
≮	not less than	:	is to

Arithmetic operations

The four basic arithmetic operations are addition, subtraction, multiplication and division. Each part of an arithmetic operation has a specific name.

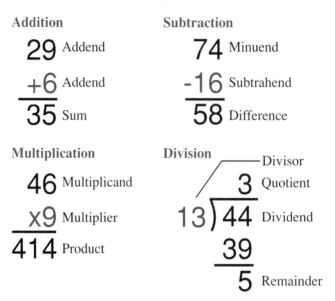

Addition

29 Addend

+6 Addend

35 Sum

Subtraction

74 Minuend

-16 Subtrahend

58 Difference

Multiplication

46 Multiplicand

x9 Multiplier

414 Product

Division

Divisor

3 Quotient

13)44 Dividend

39

5 Remainder

Fraction

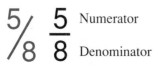

5/8   5/8 Numerator / Denominator

Simple (or vulgar) fraction

9/7   9/7 Numerator / Denominator

Binary numbers
The binary system is formulated on a base of 2, or on a sum of powers of 2. For example, the number 101011 is equal to $2^5 + 0 + 2^3 + 0 + 2^1 + 2^0$; in the decimal system, this number equals 43. The system is used frequently in computer applications.

In describing computer storage, 1 bit = 1 binary digit; 1 byte = 8 bits in most systems; 1 megabyte (MB) = 1 048 576 bytes. The table below shows other decimal/binary equivalents.

Decimal	Binary	Decimal	Binary
1	1	21	10101
2	10	30	11110
3	11	40	101000
4	100	50	110010
5	101	60	111100
6	110	90	1011010
7	111	100	1100100
8	1000	200	11001000
9	1001	300	100101100
10	1010	400	110010000
11	1011	500	111110100
12	1100	600	1001011000
13	1101	900	1110000100
14	1110	1 000	1111101000
15	1111	2 000	11111010000
16	10000	4 000	111110100000
17	10001	5 000	1001110001000
18	10010	10 000	10011100010000
19	10011	20 000	100111000100000
20	10100	100 000	11000011010100000

Computer coding systems

ASCII (American Standard Code for Information Interchange) is an international coding system of character representation. Its 256 codes represent computer commands and letters of the alphabet. Hexadecimal is a system of numbering based on 16 digits (as opposed to 10 in the decimal system): 1 to 9 and A to F.

Binary, ASCII and hexadecimal systems are used in computer programming.

The table below shows character equivalents in decimal, hexadecimal and ASCII systems.

Dec	Hex	ASCII	Dec	Hex	ASCII
000	00	NUL	016	10	DLE
001	01	SOH	017	11	DC1
002	02	STX	018	12	DC2
003	03	ETX	019	13	DC3
004	04	EOT	020	14	DC4
005	05	ENQ	021	15	NAK
006	06	ACK	022	16	SYN
007	07	BEL	023	17	ETB
008	08	BS	024	18	CAN
009	09	HT	025	19	EM
010	0A	LF	026	1A	SUB
011	0B	VT	027	1B	ESCAPE
012	0C	FF	028	1C	FS
013	0D	CR	029	1D	GS
014	0E	SO	030	1E	RS
015	0F	SI	031	1F	US

Dec	Hex	ASCII
032	20	SPACE
033	21	!
034	22	"
035	23	#
036	24	$
037	25	%
038	26	&
039	27	'
040	28	(
041	29	)
042	2A	*
043	2B	+
044	2C	,
045	2D	–
046	2E	.
047	2F	/
048	30	0
049	31	1
050	32	2
051	33	3
052	34	4
053	35	5
054	36	6
055	37	7
056	38	8

Dec	Hex	ASCII
057	39	9
058	3A	:
059	3B	;
060	3C	<
061	3D	=
062	3E	>
063	3F	?
064	40	@
065	41	A
066	42	B
067	43	C
068	44	D
069	45	E
070	46	F
071	47	G
072	48	H
073	49	I
074	4A	J
075	4B	K
076	4C	L
077	4D	M
078	4E	N
079	4F	O
080	50	P
081	51	Q

Dec	Hex	ASCII
082	52	R
083	53	S
084	54	T
085	55	U
086	56	V
087	57	W
088	58	X
089	59	Y
090	5A	Z
091	5B	[
092	5C	\
093	5D	]
094	5E	^
095	5F	_
096	60	`
097	61	a
098	62	b
099	63	c
100	64	d
101	65	e
102	66	f
103	67	g
104	68	h
105	69	i
106	6A	j

Dec	Hex	ASCII
107	6B	k
108	6C	l
109	6D	m
110	6E	n
111	6F	o
112	70	p
113	71	q
114	72	r
115	73	s
116	74	t
117	75	u
118	76	v
119	77	w
120	78	x
121	79	y
122	7A	z
123	7B	{
124	7C		
125	7D	}
126	7E	~
127	7F	DEL

Fractions, decimals and percentages

Fraction			Decimal	Percentage
$1/9$			0.111 111	11.11%
$1/7$			0.142 857	14.29%
$1/6$			0.166 667	16.67%
$1/5$			0.2	20.00%
$2/9$			0.222 222	22.22%
$2/7$			0.285 714	28.58%
$3/9$	$2/6$	$1/3$	0.333 333	33.33%
$2/5$			0.4	40.00%
$3/7$			0.428 571	42.86%
$4/9$			0.444 444	44.44%
$3/6$			0.5	50.00%
$5/9$			0.555 555	55.56%
$4/7$			0.571 429	57.14%
$3/5$			0.6	60.00%
$6/9$	$4/6$	$2/3$	0.666 666	66.67%
$5/7$			0.714 286	71.43%
$7/9$			0.777 778	77.78%
$4/5$			0.8	80.00%
$5/6$			0.833 333	83.33%
$6/7$			0.857 143	85.71%
$8/9$			0.888 889	88.89%
$9/9$ $7/7$ $6/6$ $5/5$ $3/3$			1	100%

Fraction					Decimal	Percentage
$^1/_{64}$					0.015 625	1.56%
$^2/_{64}$	$^1/_{32}$				0.031 25	3.13%
$^3/_{64}$					0.046 875	4.69%
$^4/_{64}$	$^2/_{32}$	$^1/_{16}$			0.062 5	6.25%
$^5/_{64}$					0.078 125	7.81%
$^6/_{64}$	$^3/_{32}$				0.093 75	9.38%
$^7/_{64}$					0.109 375	10.94%
$^8/_{64}$	$^4/_{32}$	$^2/_{16}$	$^1/_8$		0.125	12.50%
$^9/_{64}$					0.140 625	14.06%
$^{10}/_{64}$	$^5/_{32}$				0.156 25	15.63%
$^{11}/_{64}$					0.171 875	17.19%
$^{12}/_{64}$	$^6/_{32}$	$^3/_{16}$			0.187 5	18.75%
$^{13}/_{64}$					0.203 125	20.31%
$^{14}/_{64}$	$^7/_{32}$				0.218 75	21.88%
$^{15}/_{64}$					0.234 375	23.44%
$^{16}/_{64}$	$^8/_{32}$	$^4/_{16}$	$^2/_8$	$^1/_4$	0.25	25.00%
$^{17}/_{64}$					0.265 625	26.56%
$^{18}/_{64}$	$^9/_{32}$				0.281 25	28.13%
$^{19}/_{64}$					0.296 875	29.69%
$^{20}/_{64}$	$^{10}/_{32}$	$^5/_{16}$			0.312 5	31.25%
$^{21}/_{64}$					0.328 125	32.81%
$^{22}/_{64}$	$^{11}/_{32}$				0.343 75	34.38%

Fractions, decimals and percentages (continued)

Fraction	Decimal	Percentage
23/64	0.359 375	35.94%
24/64 12/32 6/16 3/8	0.375	37.50%
25/64	0.390 625	39.06%
26/64 12/32	0.406 25	40.63%
27/64	0.421 875	42.19%
28/64 14/32 7/16	0.437 5	43.75%
29/64	0.453 125	45.31%
30/64 15/32	0.468 75	46.88%
31/64	0.484 375	48.44%
32/64 16/32 8/16 4/8 2/4 1/2	0.5	50.00%
33/64	0.515 625	51.56%
34/64 17/32	0.531 25	53.13%
35/64	0.546 875	54.69%
36/64 18/32 9/16	0.562 5	56.25%
37/64	0.578 125	57.81%
38/64 19/32	0.593 75	59.37%
39/64	0.609 375	60.94%
40/64 20/32 10/16 5/8	0.625	62.50%
41/64	0.640 625	64.06%
42/64 21/32	0.656 25	65.63%
43/64	0.671 875	67.19%
44/64 22/32 11/16	0.687 5	68.75%

Fraction	Decimal	Percentage
45/64	0.703 125	70.31%
46/64 23/32	0.718 75	71.88%
47/64	0.734 375	73.44%
48/64 24/32 12/16 6/8 3/4	0.75	75.00%
49/64	0.765 625	76.56%
50/64 25/32	0.781 25	78.13%
51/64	0.796 875	79.69%
52/64 26/32 13/16	0.812 5	81.25%
53/64	0.828 125	82.81%
54/64 27/32	0.843 75	84.38%
55/64	0.859 375	85.94%
56/64 28/32 14/16 7/8	0.875	87.50%
57/64	0.890 625	89.06%
58/64 29/32	0.906 25	90.63%
59/64	0.921 875	92.19%
60/64 30/32 15/16	0.937 5	93.75%
61/64	0.953 125	95.31%
62/64 31/32	0.968 75	96.88%
63/64	0.984 375	98.44%
64/64 32/32 16/16 8/8 4/4 2/2	1	100%

Prime numbers
These are whole numbers that have only two factors –
the number itself and the number 1. The only even
prime number is 2: all other prime numbers are odd.

There are an infinite number of prime numbers. The
first 126 are given below. The number at the foot of the
table is the largest prime known in 1952. The largest
prime known in 1983 has 39 751 digits.

2	47	109	191	269	353	439	523	617
3	53	113	193	271	359	443	541	619
5	59	127	197	277	367	449	547	631
7	61	131	199	281	373	457	557	641
11	67	137	211	283	379	461	563	643
13	71	139	223	293	383	463	569	647
17	73	149	227	307	389	467	571	653
19	79	151	229	311	397	479	577	659
23	83	157	233	313	401	487	587	661
29	89	163	239	317	409	491	593	673
31	97	167	241	331	419	499	599	677
37	101	173	251	337	421	503	601	683
41	103	179	257	347	431	509	607	691
43	107	181	263	349	433	521	613	701

170141183460469231731687303715884105727

Fibonacci sequence

Each number in a Fibonacci sequence is the sum of the two numbers preceding it. The sequence can therefore be built up using simple addition. Below is an example of a Fibonacci sequence.

0 + 1 = **2**	1597 + 987 = **2584**
2 + 1 = **3**	2584 + 1597 = **4181**
3 + 2 = **5**	4181 + 2584 = **6765**
5 + 3 = **8**	6765 + 4181 = **10 946**
8 + 5 = **13**	10 946 + 6765 = **17 711**
13 + 8 = **21**	17 711 + 10 946 = **28 657**
21 + 13 = **34**	28 657 + 17 711 = **46 368**
34 + 21 = **55**	46 368 + 28 657 = **75 025**
55 + 34 = **89**	75 025 + 46 368 = **121 393**
89 + 55 = **144**	121 393 + 75 025 = **196 418**
144 + 89 = **233**	196 418 + 121 393 = **317 811**
233 + 144 = **377**	317 811 + 196 418 = **514 229**
377 + 233 = **610**	514 229 + 317 811 = **832 040**
610 + 377 = **987**	832 040 + 514 229 = **1 346 269**

Square and cube roots
*Accurate to 3 decimal places – they have not been rounded up or down.

Square and cube* roots of 1 to 25			Square and cube* roots of 26 to 50		
	$\sqrt{}$	$\sqrt[3]{}$		$\sqrt{}$	$\sqrt[3]{}$
1	1.000	1.000	26	5.099	2.962
2	1.414	1.259	27	5.196	3.000
3	1.732	1.442	28	5.291	3.036
4	2.000	1.587	29	5.385	3.072
5	2.236	1.709	30	5.477	3.107
6	2.449	1.817	31	5.567	3.141
7	2.645	1.912	32	5.656	3.174
8	2.828	2.000	33	5.744	3.207
9	3.000	2.080	34	5.831	3.239
10	3.162	2.154	35	5.916	3.271
11	3.316	2.223	36	6.000	3.301
12	3.464	2.289	37	6.082	3.332
13	3.605	2.351	38	6.614	3.361
14	3.741	2.410	39	6.245	3.391
15	3.873	2.466	40	6.324	3.419
16	4.000	2.519	41	6.403	3.448
17	4.123	2.571	42	6.480	3.476
18	4.242	2.620	43	6.557	3.503
19	4.358	2.668	44	6.633	3.530
20	4.472	2.714	45	6.708	3.556
21	4.582	2.758	46	6.782	3.583
22	4.690	2.802	47	6.855	3.608
23	4.795	2.843	48	6.928	3.634
24	4.899	2.884	49	7.000	3.659
25	5.000	2.924	50	7.071	3.684

Square and cube* roots of 51 to 75		
	$\sqrt{}$	$\sqrt[3]{}$
51	7.141	3.708
52	7.211	3.732
53	7.280	3.756
54	7.348	3.779
55	7.416	3.802
56	7.483	3.825
57	7.549	3.848
58	7.615	3.870
59	7.681	3.893
60	7.746	3.913
61	7.810	3.936
62	7.874	3.957
63	7.937	3.979
64	8.000	4.000
65	8.062	4.020
66	8.124	4.041
67	8.185	4.061
68	8.246	4.081
69	8.306	4.101
70	8.366	4.121
71	8.426	4.140
72	8.485	4.160
73	8.544	4.179
74	8.602	4.198
75	8.660	4.217

Square and cube* roots of 76 to 100		
	$\sqrt{}$	$\sqrt[3]{}$
76	8.717	4.235
77	8.775	4.254
78	8.831	4.272
79	8.888	4.290
80	8.944	4.308
81	9.000	4.326
82	9.055	4.344
83	9.110	4.362
84	9.165	4.379
85	9.219	4.396
86	9.273	4.414
87	9.327	4.431
88	9.380	4.447
89	9.434	4.464
90	9.486	4.481
91	9.539	4.497
92	9.591	4.514
93	9.643	4.530
94	9.695	4.546
95	9.746	4.562
96	9.798	4.578
97	9.848	4.594
98	9.899	4.610
99	9.949	4.626
100	10.00	4.641

Multiplication tables

×2		×3		×4		×5		×6	
1	2	1	3	1	4	1	5	1	6
2	4	2	6	2	8	2	10	2	12
3	6	3	9	3	12	3	15	3	18
4	8	4	12	4	16	4	20	4	24
5	10	5	15	5	20	5	25	5	30
6	12	6	18	6	24	6	30	6	36
7	14	7	21	7	28	7	35	7	42
8	16	8	24	8	32	8	40	8	48
9	18	9	27	9	36	9	45	9	54
10	20	10	30	10	40	10	50	10	60
11	22	11	33	11	44	11	55	11	66
12	24	12	36	12	48	12	60	12	72
13	26	13	39	13	52	13	65	13	78
14	28	14	42	14	56	14	70	14	84
15	30	15	45	15	60	15	75	15	90
16	32	16	48	16	64	16	80	16	96
17	34	17	51	17	68	17	85	17	102
18	36	18	54	18	72	18	90	18	108
19	38	19	57	19	76	19	95	19	114
25	50	25	75	25	100	25	125	25	150
35	70	35	105	35	140	35	175	35	210
45	90	45	135	45	180	45	225	45	270
55	110	55	165	55	220	55	275	55	330
65	130	65	195	65	260	65	325	65	390
75	150	75	225	75	300	75	375	75	450
85	170	85	255	85	340	85	425	85	510
95	190	95	285	95	380	95	475	95	570

×7		×8		×9		×10		×11	
1	7	1	8	1	9	1	10	1	11
2	14	2	16	2	18	2	20	2	22
3	21	3	24	3	27	3	30	3	33
4	28	4	32	4	36	4	40	4	44
5	35	5	40	5	45	5	50	5	55
6	42	6	48	6	54	6	60	6	66
7	49	7	56	7	63	7	70	7	77
8	56	8	64	8	72	8	80	8	88
9	63	9	72	9	81	9	90	9	99
10	70	10	80	10	90	10	100	10	110
11	77	11	88	11	99	11	110	11	121
12	84	12	96	12	108	12	120	12	132
13	91	13	104	13	117	13	130	13	143
14	98	14	112	14	126	14	140	14	154
15	105	15	120	15	135	15	150	15	165
16	112	16	128	16	144	16	160	16	176
17	119	17	136	17	153	17	170	17	187
18	126	18	144	18	162	18	180	18	198
19	133	19	152	19	171	19	190	19	209
25	175	25	200	25	225	25	250	25	275
35	245	35	280	35	315	35	350	35	385
45	315	45	360	45	405	45	450	45	495
55	385	55	440	55	495	55	550	55	605
65	455	65	520	65	585	65	650	65	715
75	525	75	600	75	675	75	750	75	825
85	595	85	680	85	765	85	850	85	935
95	665	95	760	95	855	95	950	95	1045

Multiplication tables (continued)

×12		×13		×14		×15		×16	
1	12	1	13	1	14	1	15	1	16
2	24	2	26	2	28	2	30	2	32
3	36	3	39	3	42	3	45	3	48
4	48	4	52	4	56	4	60	4	64
5	60	5	65	5	70	5	75	5	80
6	72	6	78	6	84	6	90	6	96
7	84	7	91	7	98	7	105	7	112
8	96	8	104	8	112	8	120	8	128
9	108	9	117	9	126	9	135	9	144
10	120	10	130	10	140	10	150	10	160
11	132	11	143	11	154	11	165	11	176
12	144	12	156	12	168	12	180	12	192
13	156	13	169	13	182	13	195	13	208
14	168	14	182	14	196	14	210	14	224
15	180	15	195	15	210	15	225	15	240
16	192	16	208	16	224	16	240	16	256
17	204	17	221	17	238	17	255	17	272
18	216	18	234	18	252	18	270	18	288
19	228	19	247	19	266	19	285	19	304
25	300	25	325	25	350	25	375	25	400
35	420	35	455	35	490	35	525	35	560
45	540	45	585	45	630	45	675	45	720
55	660	55	715	55	770	55	825	55	880
65	780	65	845	65	910	65	975	65	1040
75	900	75	975	75	1050	75	1125	75	1200
85	1020	85	1105	85	1190	85	1275	85	1360
95	1140	95	1235	95	1330	95	1425	95	1520

×17		×18		×19		×20		×21	
1	17	1	18	1	19	1	20	1	21
2	34	2	36	2	38	2	40	2	42
3	51	3	54	3	57	3	60	3	63
4	68	4	72	4	76	4	80	4	84
5	85	5	90	5	95	5	100	5	105
6	102	6	108	6	114	6	120	6	126
7	119	7	126	7	133	7	140	7	147
8	136	8	144	8	152	8	160	8	168
9	153	9	162	9	171	9	180	9	189
10	170	10	180	10	190	10	200	10	210
11	187	11	198	11	209	11	220	11	231
12	204	12	216	12	228	12	240	12	252
13	221	13	234	13	247	13	260	13	273
14	238	14	252	14	266	14	280	14	294
15	255	15	270	15	285	15	300	15	315
16	272	16	288	16	304	16	320	16	336
17	289	17	306	17	323	17	340	17	357
18	306	18	324	18	342	18	360	18	378
19	323	19	342	19	361	19	380	19	399
25	425	25	450	25	475	25	500	25	525
35	595	35	630	35	665	35	700	35	735
45	765	45	810	45	855	45	900	45	945
55	935	55	990	55	1045	55	1100	55	1155
65	1105	65	1170	65	1235	65	1300	65	1365
75	1275	75	1350	75	1425	75	1500	75	1575
85	1445	85	1530	85	1615	85	1700	85	1785
95	1615	95	1710	95	1805	95	1900	95	1995

Multiplication grid

Below is a quick-reference grid giving products and quotients. It can be used for either multiplication or division.

	Column											
Row	**1**	**2**	**3**	**4**	**5**	**6**	**7**	**8**	**9**	**10**	**11**	**12**
1	1	2	3	4	5	6	7	8	9	10	11	12
2	2	4	6	8	10	12	14	16	18	20	22	24
3	3	6	9	12	15	18	21	24	27	30	33	36
4	4	8	12	16	20	24	28	32	36	40	44	48
5	5	10	15	20	25	30	35	40	45	50	55	60
6	6	12	18	24	30	36	42	48	54	60	66	72
7	7	14	21	28	35	42	49	56	63	70	77	84
8	8	16	24	32	40	48	56	64	72	80	88	96
9	9	18	27	36	45	54	63	72	81	90	99	108
10	10	20	30	40	50	60	70	80	90	100	110	120
11	11	22	33	44	55	66	77	88	99	110	121	132
12	12	24	36	48	60	72	84	96	108	120	132	144

Multiplication

To multiply 6 by 9, for example, scan down column six until you reach row nine. The number in the square where column six intersects row nine is the product, 54.

Division

To divide 56 by 8, scan down column eight to find 56 (the dividend) then scan across to find the row number. This is the quotient, 7.

Interest
Interest refers to the charge made for borrowing money.
It is usually expressed in terms of percentage rates.
There are two types of interest: simple interest and
compound interest.

Simple interest
This type of interest is calculated on the amount of
money originally loaned (the principal). The formula
used to calculate simple interest is:

$$I = \frac{P \times R \times T}{100}$$

I is interest, P is principal, R is the percentage rate per
unit time, and T is the length of time (measured in
units) over which the money is invested or loaned.
   The final sum – or amount of money to which the
principal will grow – is figured using the formula:

$$S \text{ (sum)} = P\left(1 + \frac{R \times T}{100}\right)$$

Compound interest
Unlike simple interest, which is paid only on the
principal, compound interest is paid also on the
previous interest earned. Thus the sum – or amount to
which the principal will grow – increases at a much
faster rate than with simple interest.
   Compound interest is figured using the formula:

$$S = P(1 + i)^n$$

The 'i' represents the periodic interest; 'n' is the
number of periods.

Simple interest

Simple interest (in pounds) to add to £1000 (percent per annum)				
	2.5%	**3%**	**3.5%**	**4%**
1 day	0.069	0.083	0.097	0.111
2 days	0.139	0.167	0.194	0.222
3 days	0.208	0.250	0.292	0.333
4 days	0.278	0.333	0.389	0.444
5 days	0.347	0.417	0.486	0.556
6 days	0.417	0.500	0.583	0.667
30 days	2.083	2.500	2.917	3.333
60 days	4.167	5.000	5.833	6.667
90 days	6.250	7.500	8.750	10.000
180 days	12.500	15.000	17.500	20.000
360 days	25.000	30.000	35.000	40.000

Simple interest (in pounds) added on to a principal of £100 (percent per annum)				
	7%	**8%**	**9%**	**10%**
1 year	107 00	108 00	109 00	110 00
5 years	135 00	140 00	145 00	150 00
10 years	170 00	180 00	190 00	200 00
20 years	240 00	260 00	280 00	300 00
30 years	310 00	340 00	370 00	400 00
40 years	380 00	420 00	460 00	500 00
50 years	450 00	500 00	550 00	600 00

4.5%	5%	5.5%	6%	6.5%	7%
0.125	0.139	0.153	0.167	0.185	0.194
0.250	0.278	0.306	0.333	0.361	0.389
0.375	0.417	0.458	0.500	0.545	0.583
0.500	0.556	0.611	0.667	0.722	0.778
0.625	0.694	0.764	0.833	0.903	0.972
0.750	0.833	0.917	1.000	1.083	1.167
3.750	4.167	4.583	5.000	5.417	5.833
7.500	8.333	9.167	10.000	10.833	11.667
11.250	12.500	13.750	15.000	16.250	17.500
22.500	25.000	27.500	30.000	32.500	35.000
45.000	50.000	55.000	60.000	65.000	70.000

11%	12%	13%	14%	15%
111 00	112 00	113 00	114 00	115 00
155 00	160 00	165 00	170 00	175 00
210 00	220 00	230 00	240 00	250 00
320 00	340 00	360 00	380 00	400 00
430 00	460 00	490 00	520 00	550 00
540 00	580 00	620 00	660 00	700 00
650 00	700 00	750 00	800 00	850 00

Compound interest

The table below shows the compound interest paid (in pounds) on a principal of £100. The interest rate is in percent per annum.

Period	4%	5%	6%	7%
1 day	0.011	0.014	0.016	0.019
1 week	0.077	0.096	0.115	0.134
6 months	2.00	2.50	3.00	3.50
1 year	4.00	5.00	6.00	7.00
2 years	8.16	10.25	12.36	14.49
3 years	12.49	15.76	19.10	22.50
4 years	16.99	21.55	26.25	31.08
5 years	21.67	27.63	33.82	40.26
6 years	26.53	34.01	41.85	50.07
7 years	31.59	40.71	50.36	60.58
8 years	36.86	47.75	59.38	71.82
9 years	42.33	55.13	68.95	83.85
10 years	48.02	62.89	79.08	96.72

Comparing the two

Money grows much more quickly with compound interest than with simple interest. Compare, for example, the amount of time required for an amount of money to double itself with simple interest and with compound interest:

8%	9%	10%	12%	14%	16%
0.022	0.025	0.027	0.033	0.038	0.044
0.153	0.173	0.192	0.230	0.268	0.307
4.00	4.50	5.00	6.00	7.00	8.00
8.00	9.00	10.00	12.00	14.00	16.00
16.64	18.81	21.00	25.44	29.96	34.56
25.97	29.50	33.10	40.49	48.15	56.09
36.05	41.16	46.41	57.35	68.90	81.06
46.93	53.86	61.05	76.23	92.54	110.03
58.69	67.71	77.16	97.38	119.50	143.64
71.38	82.80	94.87	121.07	150.23	182.62
85.09	99.26	114.36	147.60	185.26	227.84
99.90	117.19	135.79	177.31	225.19	280.30
115.89	136.74	159.37	210.58	270.72	341.14

Rate	Simple	Compound
7%	14 yrs, 104 days	10 yrs, 89 days
10%	10 yrs	7 yrs, 100 days

2: Length and area

Formulas: length
Below are listed the multiplication/division factors for
converting units of length from imperial to metric, and
vice versa. Note that two kinds of factors are given:
quick, for an approximate conversion that can be made
without a calculator; and accurate, for an exact
conversion.

Milli-inches (mils) Micrometres (μm)		**Quick**	**Accurate**
mils ⟶ μm		× 25	× 25.4
μm ⟶ mils		÷ 25	× 0.0394
Inches (in) Millimetres (mm)			
in ⟶ mm		× 25	× 25.4
mm ⟶ in		÷ 25	× 0.0394
Inches (in) Centimetres (cm)			
in ⟶ cm		× 2.5	× 2.54
cm ⟶ in		÷ 2.5	× 0.394
Feet (ft) Metres (m)			
ft ⟶ m		÷ 3.3	× 0.305
m ⟶ ft		× 3.3	× 3.281
Yards (yd) Metres (m)			
yd ⟶ m		÷ 1	× 0.914
m ⟶ yd		× 1	× 1.094

Fathoms (fm) Metres (m)

		Quick	**Accurate**
fm ⟶ m		× 2	× 1.83
m ⟶ fm		÷ 2	× 0.547

Chains (ch) Metres (m)

		Quick	Accurate
ch ⟶ m		× 20	× 20.108
m ⟶ ch		÷ 20	× 0.0497

Furlongs (fur) Metres (m)

		Quick	Accurate
fur ⟶ m		× 200	× 201.17
m ⟶ fur		÷ 200	× 0.005

Yards (yd) Kilometres (km)

		Quick	Accurate
yd ⟶ km		÷ 1000	× 0.00091
km ⟶ yd		× 1000	× 1093.6

Miles (mi) Kilometres (km)

		Quick	Accurate
mi ⟶ km		× 1.5	× 1.609
km ⟶ mi		÷ 1.5	× 0.621

Nautical miles (n mi) Miles (mi)

		Quick	Accurate
n mi ⟶ mi		× 1.2	× 1.151
mi ⟶ n mi		÷ 1.2	× 0.869

Nautical miles (n mi) Kilometres (km)

		Quick	Accurate
n mi ⟶ km		× 2	× 1.852
km ⟶ n mi		÷ 2	× 0.54

Conversion tables: length

The tables below can be used to convert units of length from one measuring system to another. The first group of tables converts imperial to metric; the second, beginning on page 66, converts metric to imperial.

Milli-inches to Micrometres		Inches to Millimetres		Inches to Centimetres	
mils	μm	in	mm	in	cm
1	25.4	1	25.4	1	2.54
2	50.8	2	50.8	2	5.08
3	76.2	3	76.2	3	7.62
4	101.6	4	101.6	4	10.16
5	127.0	5	127.0	5	12.70
6	152.4	6	152.4	6	15.24
7	177.8	7	177.8	7	17.78
8	203.2	8	203.2	8	20.32
9	228.6	9	228.6	9	22.86
10	254.0	10	254.0	10	25.40
20	508.0	20	508.0	20	50.80
30	762.0	30	762.0	30	76.20
40	1016.0	40	1016.0	40	101.60
50	1270.0	50	1270.0	50	127.00
60	1524.0	60	1524.0	60	152.40
70	1778.0	70	1778.0	70	177.80
80	2032.0	80	2032.0	80	203.20
90	2286.0	90	2286.0	90	228.60
100	2540.0	100	2540.0	100	254.00

Feet to Metres		Yards to Metres		Fathoms to Metres	
ft	**m**	**yd**	**m**	**fm**	**m**
1	0.305	1	0.914	1	1.83
2	0.610	2	1.829	2	3.66
3	0.914	3	2.743	3	5.49
4	1.219	4	3.658	4	7.32
5	1.524	5	4.572	5	9.14
6	1.829	6	5.486	6	10.97
7	2.134	7	6.401	7	12.80
8	2.438	8	7.315	8	14.63
9	2.743	9	8.230	9	16.46
10	3.048	10	9.144	10	18.29
20	6.096	20	18.288	20	36.58
30	9.144	30	27.432	30	54.87
40	12.192	40	36.576	40	73.16
50	15.240	50	45.720	50	91.45
60	18.288	60	54.864	60	109.74
70	21.336	70	64.008	70	128.03
80	24.384	80	73.152	80	146.32
90	27.432	90	82.296	90	164.61
100	30.480	100	91.440	100	182.90

Imperial and metric units of length (continued)

Chains to Metres		Furlongs to Metres		Yards to Kilometres	
ch	m	fur	m	yd	km
1	20.108	1	201.17	100	0.091
2	40.216	2	402.34	200	0.183
3	60.324	3	603.50	300	0.274
4	80.432	4	804.67	400	0.366
5	100.540	5	1005.84	500	0.457
6	120.648	6	1207.01	600	0.549
7	140.756	7	1408.18	700	0.640
8	160.864	8	1609.34	800	0.731
9	180.972	9	1810.51	900	0.823
10	201.080	10	2011.68	1000	0.914
20	402.160	20	4023.36	2000	1.829
30	603.240	30	6035.04	3000	2.743
40	804.320	40	8046.72	4000	3.658
50	1005.400	50	10 058.40	5000	4.572
60	1206.480	60	12 070.08	6000	5.486
70	1407.560	70	14 081.76	7000	6.401
80	1608.640	80	16 093.44	8000	7.315
90	1809.720	90	18 105.12	9000	8.230
100	2010.800	100	20 116.80	10 000	9.144

Miles to Kilometres		Nautical miles to Miles		Nautical miles to Kilometres	
mi	km	n mi	mi	n mi	km
1	1.609	1	1.151	1	1.852
2	3.219	2	2.302	2	3.704
3	4.828	3	3.452	3	5.556
4	6.437	4	4.603	4	7.408
5	8.047	5	5.754	5	9.260
6	9.656	6	6.905	6	11.112
7	11.265	7	8.055	7	12.964
8	12.875	8	9.206	8	14.816
9	14.484	9	10.357	9	16.668
10	16.093	10	11.508	10	18.520
20	32.187	20	23.016	20	37.040
30	48.280	30	34.523	30	55.560
40	64.374	40	46.031	40	74.080
50	80.467	50	57.539	50	92.600
60	96.561	60	69.047	60	111.120
70	112.654	70	80.554	70	129.640
80	128.748	80	92.062	80	148.160
90	144.841	90	103.570	90	166.680
100	160.934	100	115.078	100	185.200

Imperial and metric units of length (continued)

Micrometres to Milli-inches		Millimetres to Inches		Centimetres to Inches	
μm	mils	mm	in	cm	in
1	0.039	1	0.039	1	0.394
2	0.079	2	0.079	2	0.787
3	0.118	3	0.118	3	1.181
4	0.157	4	0.157	4	1.575
5	0.197	5	0.197	5	1.969
6	0.236	6	0.236	6	2.362
7	0.276	7	0.276	7	2.756
8	0.315	8	0.315	8	3.150
9	0.354	9	0.354	9	3.543
10	0.394	10	0.394	10	3.937
20	0.787	20	0.787	20	7.874
30	1.181	30	1.181	30	11.811
40	1.575	40	1.575	40	15.748
50	1.969	50	1.969	50	19.685
60	2.362	60	2.362	60	23.622
70	2.756	70	2.756	70	27.559
80	3.150	80	3.150	80	31.496
90	3.543	90	3.543	90	35.433
100	3.937	100	3.937	100	39.370

Metres to Feet	
m	ft
1	3.281
2	6.562
3	9.843
4	13.123
5	16.404
6	19.685
7	22.966
8	26.247
9	29.528
10	32.808
20	65.617
30	98.425
40	131.234
50	164.042
60	196.850
70	229.659
80	262.467
90	295.276
100	328.084

Metres to Yards	
m	yd
1	1.094
2	2.187
3	3.281
4	4.374
5	5.468
6	6.562
7	7.655
8	8.749
9	9.843
10	10.936
20	21.872
30	32.808
40	43.745
50	54.681
60	65.617
70	76.553
80	87.489
90	98.425
100	109.361

Metres to Fathoms	
m	fm
1	0.547
2	1.093
3	1.640
4	2.187
5	2.734
6	3.280
7	3.827
8	4.374
9	4.921
10	5.467
20	10.935
30	16.402
40	21.870
50	27.337
60	32.805
70	38.272
80	43.740
90	49.207
100	54.674

Imperial and metric units of length (continued)

Metres to Chains		Metres to Furlongs		Kilometres to Yards	
m	ch	m	fur	km	yd
1	0.0497	1	0.005	1	1093.6
2	0.0994	2	0.010	2	2187.2
3	0.1491	3	0.015	3	3280.8
4	0.1989	4	0.020	4	4374.4
5	0.2487	5	0.025	5	5468.0
6	0.2983	6	0.030	6	6561.6
7	0.3481	7	0.035	7	7655.2
8	0.3979	8	0.040	8	8748.8
9	0.4476	9	0.045	9	9842.4
10	0.4973	10	0.050	10	10 936.0
20	0.9946	20	0.099	20	21 872.0
30	1.4919	30	0.149	30	32 808.0
40	1.9893	40	0.199	40	43 744.0
50	2.4866	50	0.249	50	54 680.0
60	2.9839	60	0.298	60	65 616.0
70	3.4812	70	0.348	70	76 552.0
80	3.9785	80	0.398	80	87 488.0
90	4.4758	90	0.447	90	98 424.0
100	4.9731	100	0.497	100	109 360.0

Kilometres to Miles		Miles to Nautical miles		Kilometres to Nautical miles	
km	mi	mi	n mi	km	n mi
1	0.621	1	0.869	1	0.54
2	1.243	2	1.738	2	1.08
3	1.864	3	2.607	3	1.62
4	2.485	4	3.476	4	2.16
5	3.107	5	4.349	5	2.70
6	3.728	6	5.214	6	3.24
7	4.350	7	6.083	7	3.78
8	4.971	8	6.952	8	4.32
9	5.592	9	7.821	9	4.86
10	6.214	10	8.690	10	5.40
20	12.427	20	17.380	20	10.80
30	18.641	30	26.069	30	16.20
40	24.855	40	34.759	40	21.60
50	31.069	50	43.449	50	27.00
60	37.282	60	52.139	60	32.40
70	43.496	70	60.828	70	37.80
80	49.710	80	69.518	80	43.20
90	55.923	90	78.208	90	48.60
100	62.137	100	86.900	100	54.00

Formulas: area
Below are listed the multiplication/division factors for
converting units of area from imperial to metric, and
vice versa. Note that two kinds of factors are given:
quick, for an approximate conversion that can be made
without a calculator; and accurate, for an exact
conversion.

		Quick	Accurate
Circular mils (cmil) Square micrometres (μm^2)			
cmil $\longrightarrow$ μm^2		$\times 500$	$\times 506.7$
μm^2 $\longrightarrow$ cmil		$\div 500$	$\times 0.002$
Square inches (in^2) Square millimetres (mm^2)			
in^2 $\longrightarrow$ mm^2		$\times 650$	$\times 645.2$
mm^2 $\longrightarrow$ in^2		$\div 650$	$\times 0.0015$
Square inches (in^2) Square centimetres (cm^2)			
in^2 $\longrightarrow$ cm^2		$\times 6.5$	$\times 6.452$
cm^2 $\longrightarrow$ in^2		$\div 6.5$	$\times 0.15$
Square chains (ch^2) Square metres (m^2)			
ch^2 $\longrightarrow$ m^2		$\times 400$	$\times 404.686$
m^2 $\longrightarrow$ ch^2		$\div 400$	$\times 0.0025$

	Square miles (mi^2) Square kilometres (km^2)	**Quick**	**Accurate**
	mi^2 → km^2	× 2.5	× 2.590
	km^2 → mi^2	÷ 2.5	× 0.386
	Square miles (mi^2) Hectares (ha)		
	mi^2 → ha	× 250	× 258.999
	ha → mi^2	÷ 250	× 0.0039
	Hectares (ha) Acres		
	ha → acre	× 2.5	× 2.471
	acre → ha	÷ 2.5	× 0.405
	Square metres (m^2) Square yards (yd^2)		
	m^2 → yd^2	× 1	× 1.196
	yd^2 → m^2	÷ 1	× 0.836
	Square metres (m^2) Square feet (ft^2)		
	m^2 → ft^2	× 11	× 10.764
	ft^2 → m^2	÷ 11	× 0.093

Conversion tables: area

The tables below can be used to convert units of area from one measuring system to another. The first group of tables converts imperial to metric; the second, beginning on page 75, converts metric to imperial.

Circular mils to Square micrometres		Square inches to Square millimetres		Square inches to Square centimetres	
cmil	μm^2	in²	mm²	in²	cm²
1	506.7	1	645.2	1	6.452
2	1013.4	2	129.0	2	12.903
3	1520.1	3	193.6	3	19.355
4	2026.8	4	258.1	4	25.806
5	2533.5	5	322.6	5	32.258
6	3040.2	6	387.1	6	38.710
7	3546.9	7	451.6	7	45.161
8	4053.6	8	516.1	8	51.613
9	4560.3	9	580.6	9	58.064
10	5067.0	10	6452.0	10	64.516
20	10 134.0	20	1290.0	20	129.032
30	15 201.0	30	1936.0	30	193.548
40	20 268.0	40	2581.0	40	258.064
50	25 335.0	50	3226.0	50	322.580
60	30 402.0	60	3871.0	60	387.096
70	35 469.0	70	4516.0	70	451.612
80	40 536.0	80	5161.0	80	516.128
90	45 603.0	90	5806.0	90	580.644
100	50 670.0	100	64 520.0	100	645.160

Square feet to Square metres		Square yards to Square metres		Square chains to Square metres	
ft²	m²	yd²	m²	ch²	m²
1	0.093	1	0.836	1	404.686
2	0.186	2	1.672	2	809.372
3	0.279	3	2.508	3	1214.058
4	0.372	4	3.345	4	1618.744
5	0.465	5	4.181	5	2023.430
6	0.557	6	5.017	6	2428.116
7	0.650	7	5.853	7	2832.802
8	0.743	8	6.689	8	3237.488
9	0.836	9	7.525	9	3642.174
10	0.929	10	8.361	10	4046.860
20	1.858	20	16.723	20	8093.720
30	2.787	30	25.084	30	12 140.580
40	3.716	40	33.445	40	16 187.440
50	4.645	50	41.806	50	20 234.300
60	5.574	60	50.168	60	24 281.160
70	6.503	70	58.529	70	28 328.020
80	7.432	80	66.890	80	32 374.880
90	8.361	90	75.251	90	36 421.740
100	9.290	100	83.613	100	40 468.600

Imperial and metric units of area (continued)

acre	ha
1	0.405
2	0.809
3	1.214
4	1.619
5	2.023
6	2.428
7	2.833
8	3.237
9	3.642
10	4.047
20	8.094
30	12.141
40	16.187
50	20.234
60	24.281
70	28.328
80	32.375
90	36.422
100	40.469

Acres to Hectares

mi²	ha
1	258.999
2	517.998
3	776.997
4	1035.996
5	1294.995
6	1553.994
7	1812.993
8	2071.992
9	2330.991
10	2589.990
20	5179.980
30	7769.970
40	10 359.960
50	12 949.950
60	15 539.940
70	18 129.930
80	20 719.920
90	23 309.910
100	25 899.900

Square miles to Hectares

mi²	km²
1	2.590
2	5.180
3	7.770
4	10.360
5	12.950
6	15.540
7	18.130
8	20.720
9	23.310
10	25.900
20	51.800
30	77.700
40	103.600
50	129.499
60	155.399
70	181.299
80	207.199
90	233.099
100	258.999

Square miles to Square kilometres

Square micrometres to Circular mils		Square millimetres to Square inches		Square centimetres to Square inches	
μm²	cmil	mm²	in²	cm²	in²
1	0.002	1	0.0015	1	0.155
2	0.004	2	0.0031	2	0.310
3	0.006	3	0.0047	3	0.465
4	0.008	4	0.0062	4	0.620
5	0.010	5	0.0078	5	0.775
6	0.012	6	0.0093	6	0.930
7	0.014	7	0.0109	7	1.085
8	0.016	8	0.0124	8	1.240
9	0.018	9	0.0140	9	1.395
10	0.020	10	0.0155	10	1.550
20	0.040	20	0.0310	20	3.100
30	0.060	30	0.0465	30	4.650
40	0.080	40	0.0620	40	6.200
50	0.100	50	0.0775	50	7.750
60	0.120	60	0.0930	60	9.300
70	0.140	70	0.1085	70	10.850
80	0.160	80	0.1240	80	12.400
90	0.180	90	0.1395	90	13.950
100	0.200	100	0.1550	100	15.500

Imperial and metric units of area (continued)

Square metres to Square feet		Square metres to Square yards		Square metres to Square chains	
m²	ft²	m²	yd²	m²	ch²
1	10.764	1	1.196	1	0.002
2	21.528	2	2.392	2	0.004
3	32.292	3	3.588	3	0.006
4	43.056	4	4.784	4	0.008
5	53.820	5	5.980	5	0.010
6	64.583	6	7.176	6	0.012
7	75.347	7	8.372	7	0.014
8	86.111	8	9.568	8	0.016
9	96.875	9	10.764	9	0.018
10	107.639	10	11.960	10	0.020
20	215.278	20	23.920	20	0.040
30	322.917	30	35.880	30	0.060
40	430.556	40	47.840	40	0.080
50	538.196	50	59.800	50	0.100
60	645.835	60	71.759	60	0.120
70	753.474	70	83.719	70	0.140
80	861.113	80	95.679	80	0.160
90	968.752	90	107.639	90	0.180
100	1076.391	100	119.599	100	0.200

Hectares to Acres		Hectares to Square miles		Square kilometres to Square miles	
ha	acre	ha	mi²	km²	mi²
1	2.471	1	0.003 86	1	0.386
2	4.942	2	0.007 72	2	0.772
3	7.413	3	0.011 58	3	1.158
4	9.884	4	0.015 44	4	1.544
5	12.355	5	0.019 31	5	1.931
6	14.826	6	0.023 17	6	2.317
7	17.297	7	0.027 03	7	2.703
8	19.768	8	0.030 89	8	3.089
9	22.239	9	0.034 75	9	3.475
10	24.711	10	0.038 61	10	3.861
20	49.421	20	0.077 22	20	7.722
30	74.132	30	0.115 83	30	11.583
40	98.842	40	0.154 44	40	15.444
50	123.553	50	0.193 05	50	19.305
60	148.263	60	0.231 66	60	23.166
70	172.974	70	0.270 27	70	27.027
80	197.684	80	0.308 88	80	30.888
90	222.395	90	0.347 49	90	34.749
100	247.105	100	0.386 10	100	38.610

Geometry of area
ABBREVIATIONS
a = length of top
b = length of base
h = perpendicular height
r = length of radius

$$\pi = 3.1416$$

Circle

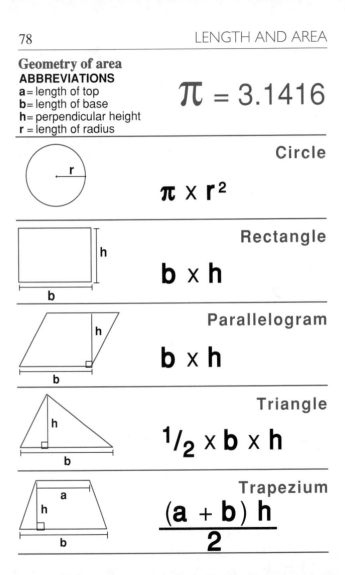

$$\pi \times r^2$$

Rectangle

$$b \times h$$

Parallelogram

$$b \times h$$

Triangle

$$\tfrac{1}{2} \times b \times h$$

Trapezium

$$\frac{(a + b)\,h}{2}$$

Geometry of surface area

ABBREVIATIONS
b = breadth of base
h = perpendicular height
l = length of base
r = length of radius

$$\pi = 3.1416$$

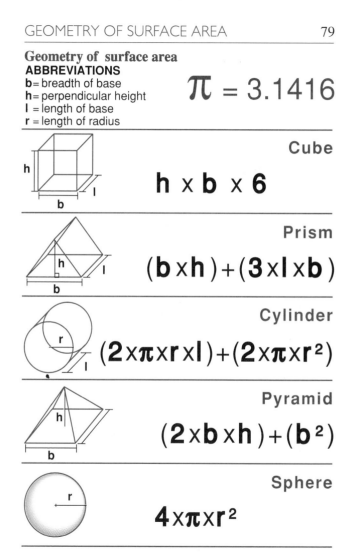

Cube

$$h \times b \times 6$$

Prism

$$(b \times h) + (3 \times l \times b)$$

Cylinder

$$(2 \times \pi \times r \times l) + (2 \times \pi \times r^2)$$

Pyramid

$$(2 \times b \times h) + (b^2)$$

Sphere

$$4 \times \pi \times r^2$$

3: Volume

Formulas

Below are listed the multiplication/division factors for converting units of volume from one measuring system to another. Note that two kinds of factors are given: quick, for an approximate conversion that can be made without a calculator; and accurate, for an exact conversion.

		Quick	**Accurate**
1 1	UK gallons (gal) US fluid gallons (fl gal)		
	UK gal ⟶ US fl gal	× 1	× 1.201
	US fl gal ⟶ UK gal	÷ 1	× 0.833
1 1	UK quarts (qt) US fluid quarts (fl qt)		
	UK qt ⟶ US fl qt	× 1	× 1.201
	US fl qt ⟶ UK qt	÷ 1	× 0.833
1 1	UK pints (pt) US fluid pints (fl pt)		
	UK pt ⟶ US fl pt	× 1	× 1.201
	US fl pt ⟶ UK pt	÷ 1	× 0.833
1 1	UK fluid ounces (fl oz) US fluid ounces (fl oz)		
	UK fl oz ⟶ US fl oz	× 1	× 0.961
	US fl oz ⟶ UK fl oz	÷ 1	× 1.041
1 2	UK fluid ounces (fl oz) Cubic inches (in³)		
	UK fl oz ⟶ in³	× 2	× 1.734
	in³ ⟶ UK fl oz	÷ 2	× 0.577

			Quick	**Accurate**
1	**16**	Cubic inches (in^3) Cubic centimetres (cm^3) in^3 ⟶ cm^3 cm^3 ⟶ in^3	× 16 ÷ 16	× 16.387 × 0.061
1	**28**	UK fluid ounces (fl oz) Millilitres (ml) UK fl oz ⟶ ml ml ⟶ UK fl oz	× 28 ÷ 28	× 28.413 × 0.035
1	**1**	UK quarts (qt) Litres (l) UK qt ⟶ l l ⟶ UK qt	× 1 ÷ 1	× 1.137 × 0.880
1	**4.5**	UK gallons (gal) Litres (l) UK gal ⟶ l l ⟶ UK gal	× 4.5 ÷ 4.5	× 4.546 × 0.220
1	**2**	Litres (l) UK pints (pt) l ⟶ UK pt UK pt ⟶ l	× 2 ÷ 2	× 1.760 × 0.568
1	**35**	Cubic metres (m^3) Cubic feet (ft^3) m^3 ⟶ ft^3 ft^3 ⟶ m^3	× 35 ÷ 35	× 35.315 × 0.028
1	**1**	Cubic metres (m^3) Cubic yards (yd^3) m^3 ⟶ yd^3 yd^3 ⟶ m^3	× 1 ÷ 1	× 1.308 × 0.765

1

Cubic metres (m³)
UK gallons (gal)   **Quick Accurate**

220

m³ ——→ UK gal   × 220   × 219.970
UK gal ——→ m³   ÷ 220   ×   0.005

1

US fluid ounces (fl oz)
Millilitres (ml)

30

US fl oz ——→ ml   × 30   × 29.572
ml ——→ US fl oz   ÷ 30   × 0.034

1

US fluid gallons (fl gal)
Litres (l)

4

US fl gal ——→ l   × 4   × 3.785
l ——→ US fl gal   ÷ 4   × 0.264

1

Litres (l)
US fluid pints (fl pt)

2

l ——→ US fl pt   × 2   × 2.113
US fl pt ——→ l   ÷ 2   × 0.473

1

Litres (l)
US fluid quarts (fl qt)

1

l ——→ US fl qt   × 1   × 1.056
US fl qt ——→ l   ÷ 1   × 0.947

1

Cubic metres (m³)
US fluid gallons (fl gal)

264

m³ ——→ US fl gal   × 264   × 264.173
US fl gal ——→ m³   ÷ 264   × 0.004

1

Cubic metres (m³)
US dry gallons (dry gal)

227

m³ ——→ dry gal   × 227   × 227.020
dry gal ——→ m³   ÷ 227   × 0.004

Conversion tables

The tables below can be used to convert units of volume from one measuring system to another. The first group of tables, beginning below, converts UK imperial to US imperial units; the second, beginning on page 85, converts US imperial to UK imperial units.

UK gallons to US fluid gallons		UK quarts to US fluid quarts	
UK gal	US fl gal	UK qt	US fl qt
1	1.201	1	1.201
2	2.402	2	2.402
3	3.603	3	3.603
4	4.804	4	4.804
5	6.005	5	6.005
6	7.206	6	7.206
7	8.407	7	8.407
8	9.608	8	9.608
9	10.809	9	10.809
10	12.010	10	12.010
20	24.020	20	24.020
30	36.030	30	36.030
40	48.040	40	48.040
50	60.050	50	60.050
60	72.060	60	72.060
70	84.070	70	84.070
80	96.080	80	96.080
90	108.090	90	108.090
100	120.100	100	120.100

UK imperial to US imperial conversions (continued)

UK pints to US fluid pints		UK fluid ounces to US fluid ounces	
UK pt	US fl pt	UK fl oz	US fl oz
1	1.201	1	0.961
2	2.402	2	1.922
3	3.603	3	2.882
4	4.804	4	3.843
5	6.005	5	4.804
6	7.206	6	5.765
7	8.407	7	6.726
8	9.608	8	7.686
9	10.809	9	8.647
10	12.010	10	9.608
20	24.020	20	19.216
30	36.030	30	28.824
40	48.040	40	38.432
50	60.050	50	48.040
60	72.060	60	57.648
70	84.070	70	67.256
80	96.080	80	76.864
90	108.090	90	86.472
100	120.100	100	96.080

US imperial to UK imperial conversions
The conversion tables below are used to convert US units of volume to UK units; tables beginning on page 93 convert US units to metric units.

US fluid gallons to UK gallons	
US fl gal	UK gal
1	0.833
2	1.665
3	2.498
4	3.331
5	4.164
6	4.998
7	5.829
8	6.662
9	7.494
10	8.327
20	16.654
30	24.981
40	33.308
50	41.635
60	49.962
70	58.289
80	66.616
90	74.943
100	83.270

US fluid quarts to UK quarts	
US fl qt	UK qt
1	0.833
2	1.665
3	2.498
4	3.331
5	4.164
6	4.996
7	5.829
8	6.662
9	7.494
10	8.327
20	16.654
30	24.981
40	33.308
50	41.635
60	49.962
70	58.289
80	66.616
90	74.943
100	83.270

US imperial to UK imperial conversions (continued)

US fluid pints to UK pints	
US fl pt	UK pt
1	0.833
2	1.665
3	2.498
4	3.331
5	4.164
6	4.996
7	5.829
8	6.662
9	7.494
10	8.327
20	16.654
30	24.981
40	33.308
50	41.635
60	49.962
70	58.289
80	66.616
90	74.943
100	83.270

US fluid ounces to UK fluid ounces	
US fl oz	UK fl oz
1	1.041
2	2.082
3	3.122
4	4.163
5	5.204
6	6.245
7	7.286
8	8.327
9	9.367
10	10.408
20	20.816
30	31.224
40	41.632
50	52.040
60	62.448
70	72.856
80	83.264
90	93.672
100	104.080

UK imperial to metric conversions

The conversion tables below are used to convert UK imperial units of volume to cubic units and metric units; tables beginning on page 90 convert metric units to UK imperial units.

UK fluid ounces to Cubic inches		Cubic inches to Cubic centimetres		Cubic feet to Cubic metres	
UK fl oz	in³	UK in³	cm³	ft³	m³
1	1.734	1	16.387	1	0.028
2	3.468	2	32.774	2	0.057
3	5.202	3	49.161	3	0.085
4	6.935	4	65.548	4	0.113
5	8.669	5	81.935	5	0.142
6	10.403	6	98.322	6	0.170
7	12.137	7	114.709	7	0.198
8	13.871	8	131.096	8	0.227
9	15.605	9	147.484	9	0.255
10	17.339	10	163.871	10	0.283
20	34.677	20	327.741	20	0.566
30	52.016	30	491.612	30	0.850
40	69.355	40	655.482	40	1.133
50	86.694	50	819.353	50	1.416
60	104.032	60	983.224	60	1.699
70	121.371	70	1147.094	70	1.982
80	138.710	80	1310.965	80	2.266
90	156.048	90	1474.835	90	2.549
100	173.387	100	1638.706	100	2.832

UK imperial to metric conversions (continued)

Cubic yards to Cubic metres		UK gallons to Cubic metres		UK gallons to Litres	
yd³	m³	UK gal	m³	UK gal	l
1	0.765	1	0.005	1	4.546
2	1.529	2	0.009	2	9.092
3	2.294	3	0.014	3	13.638
4	3.058	4	0.018	4	18.184
5	3.823	5	0.023	5	22.730
6	4.587	6	0.027	6	27.277
7	5.352	7	0.032	7	31.823
8	6.116	8	0.036	8	36.369
9	6.881	9	0.041	9	40.915
10	7.646	10	0.045	10	45.461
20	15.291	20	0.091	20	90.922
30	22.937	30	0.136	30	136.383
40	30.582	40	0.182	40	181.844
50	38.228	50	0.227	50	227.305
60	45.873	60	0.273	60	272.765
70	53.519	70	0.318	70	318.226
80	61.164	80	0.364	80	363.687
90	68.810	90	0.409	90	409.148
100	76.455	100	0.455	100	454.609

UK quarts to Litres	
UK qt	l
1	1.137
2	2.273
3	3.410
4	4.546
5	5.683
6	6.819
7	7.956
8	9.092
9	10.229
10	11.365
20	22.730
30	34.096
40	45.461
50	56.826
60	68.191
70	79.556
80	90.922
90	102.287
100	113.652

UK pints to Litres	
UK pt	l
1	0.568
2	1.137
3	1.705
4	2.273
5	2.841
6	3.410
7	3.978
8	4.546
9	5.114
10	5.683
20	11.365
30	17.048
40	22.730
50	28.413
60	34.096
70	39.778
80	45.461
90	51.143
100	56.826

UK fluid ounces to Millilitres	
UK fl oz	ml
1	28.413
2	56.826
3	85.239
4	113.652
5	142.065
6	170.478
7	198.891
8	227.305
9	255.718
10	284.131
20	568.261
30	852.392
40	1136.523
50	1420.654
60	1704.784
70	1988.915
80	2273.046
90	2557.177
100	2841.307

Metric to UK imperial conversions
The tables below convert metric units to UK imperial units.

Millilitres to UK fluid ounces		Litres to UK pints		Litres to UK quarts	
ml	UK fl oz	l	UK pt	l	UK qt
1	0.035	1	1.760	1	0.880
2	0.070	2	3.520	2	1.760
3	0.106	3	5.279	3	2.640
4	0.141	4	7.039	4	3.520
5	0.176	5	8.799	5	4.399
6	0.211	6	10.559	6	5.279
7	0.246	7	12.318	7	6.159
8	0.282	8	14.078	8	7.039
9	0.317	9	15.838	9	7.919
10	0.352	10	17.598	10	8.799
20	0.704	20	35.195	20	17.598
30	1.056	30	52.793	30	26.396
40	1.408	40	70.390	40	35.195
50	1.760	50	87.988	50	43.994
60	2.112	60	105.585	60	52.793
70	2.464	70	123.183	70	61.591
80	2.816	80	140.780	80	70.390
90	3.168	90	158.378	90	79.189
100	3.520	100	175.975	100	87.988

Litres to UK gallons	
l	UK gal
1	0.220
2	0.440
3	0.660
4	0.880
5	1.100
6	1.320
7	1.540
8	1.760
9	1.980
10	2.200
20	4.399
30	6.599
40	8.799
50	10.999
60	13.198
70	15.398
80	17.598
90	19.797
100	21.997

Cubic metres to UK gallons	
m^3	UK gal
1	219.970
2	439.940
3	659.909
4	879.879
5	1099.849
6	1319.818
7	1539.788
8	1759.757
9	1979.727
10	2199.697
20	4399.396
30	6599.093
40	8798.789
50	10 998.485
60	13 198.181
70	15 397.877
80	17 597.573
90	19 797.269
100	21 996.965

Cubic metres to Cubic feet	
m^3	ft^3
1	35.315
2	70.629
3	105.944
4	141.259
5	176.573
6	211.888
7	247.203
8	282.517
9	317.832
10	353.147
20	706.293
30	1059.440
40	1412.587
50	1765.734
60	2118.880
70	2472.027
80	2825.174
90	3178.320
100	3531.467

Metric to UK imperial conversions (continued)

Cubic metres to Cubic yards	
m³	yd³
1	1.308
2	2.616
3	3.924
4	5.232
5	6.540
6	7.848
7	9.156
8	10.464
9	11.772
10	13.080
20	26.159
30	39.239
40	52.318
50	65.398
60	78.477
70	91.557
80	104.636
90	117.716
100	130.795

Cubic centimetres to Cubic inches	
cm³	in³
1	0.061
2	0.122
3	0.183
4	0.244
5	0.305
6	0.366
7	0.427
8	0.488
9	0.549
10	0.610
20	1.220
30	1.831
40	2.441
50	3.051
60	3.661
70	4.271
80	4.882
90	5.492
100	6.102

Cubic inches to UK fluid ounces	
in³	UK fl oz
1	0.577
2	1.153
3	1.730
4	2.307
5	2.884
6	3.460
7	4.037
8	4.614
9	5.191
10	5.767
20	11.535
30	17.302
40	23.069
50	28.837
60	34.604
70	40.371
80	46.138
90	51.906
100	57.673

US imperial to metric conversions

The conversion tables below are used to convert US imperial units of volume to metric units; tables beginning on page 95 convert metric units to US imperial units.

US fluid ounces to Millilitres		US fluid pints to Litres		US fluid quarts to Litres	
US fl oz	ml	US fl pt	l	US fl qt	l
1	29.572	1	0.473	1	0.947
2	59.145	2	0.946	2	1.894
3	88.717	3	1.420	3	2.840
4	118.289	4	1.893	4	3.787
5	147.862	5	2.366	5	4.734
6	177.434	6	2.839	6	5.681
7	207.006	7	3.312	7	6.628
8	236.579	8	3.785	8	7.575
9	266.152	9	4.259	9	8.521
10	295.724	10	4.732	10	9.468
20	591.447	20	9.464	20	18.937
30	887.171	30	14.195	30	28.405
40	1182.894	40	18.927	40	37.873
50	1478.618	50	23.659	50	47.341
60	1774.341	60	28.391	60	56.810
70	2070.065	70	33.123	70	66.278
80	2365.788	80	37.854	80	75.746
90	2661.512	90	42.586	90	85.215
100	2957.235	100	47.318	100	94.683

US imperial to metric conversions (continued)

US fluid gallons to Litres		US fluid gallons to Cubic metres		US dry gallons to Cubic metres	
US fl gal	l	US fl gal	m³	US dry gal	m³
1	3.785	1	0.004	1	0.004
2	7.571	2	0.008	2	0.009
3	11.356	3	0.011	3	0.013
4	15.141	4	0.015	4	0.018
5	18.927	5	0.019	5	0.022
6	22.712	6	0.023	6	0.026
7	26.497	7	0.026	7	0.031
8	30.282	8	0.030	8	0.035
9	34.068	9	0.034	9	0.040
10	37.853	10	0.038	10	0.044
20	75.706	20	0.076	20	0.088
30	113.559	30	0.114	30	0.132
40	151.412	40	0.151	40	0.176
50	189.265	50	0.189	50	0.220
60	227.118	60	0.227	60	0.264
70	264.971	70	0.265	70	0.308
80	302.824	80	0.303	80	0.352
90	340.677	90	0.341	90	0.396
100	378.530	100	0.379	100	0.440

Metric to US imperial conversions
The tables below convert metric units to US imperial units.

Millilitres to US fluid ounces		Litres to US fluid pints		Litres to US fluid quarts	
ml	US fl oz	l	US fl pt	l	US fl qt
1	0.034	1	2.113	1	1.056
2	0.068	2	4.227	2	2.112
3	0.101	3	6.340	3	3.168
4	0.135	4	8.454	4	4.225
5	0.169	5	10.567	5	5.281
6	0.203	6	12.680	6	6.337
7	0.237	7	14.794	7	7.393
8	0.271	8	16.907	8	8.449
9	0.304	9	19.020	9	9.505
10	0.338	10	21.134	10	10.562
20	0.676	20	42.268	20	21.123
30	1.014	30	63.401	30	31.685
40	1.353	40	84.535	40	42.246
50	1.691	50	105.669	50	52.808
60	2.029	60	126.803	60	63.369
70	2.367	70	147.937	70	73.931
80	2.705	80	169.070	80	84.493
90	3.043	90	190.204	90	95.054
100	3.382	100	211.338	100	105.616

Metric to US imperial conversions (continued)

Litres to US fluid gallons		Cubic metres to US fluid gallons		Cubic metres to US dry gallons	
l	US fl gal	m³	US fl gal	m³	US dry gal
1	0.264	1	264.173	1	227.020
2	0.528	2	528.346	2	454.041
3	0.793	3	792.519	3	681.061
4	1.057	4	1056.692	4	908.081
5	1.321	5	1320.865	5	1135.102
6	1.585	6	1585.038	6	1362.122
7	1.849	7	1849.211	7	1589.143
8	2.113	8	2113.385	8	1816.163
9	2.378	9	2377.558	9	2043.183
10	2.642	10	2641.731	10	2270.204
20	5.283	20	5283.462	20	4540.407
30	7.925	30	7925.192	30	6810.611
40	10.567	40	10 566.923	40	9080.814
50	13.209	50	13 208.653	50	11 351.018
60	15.850	60	15 850.383	60	13 621.221
70	18.492	70	18 492.115	70	15 891.425
80	21.134	80	21 133.846	80	18 161.628
90	23.775	90	23 775.578	90	20 431.832
100	26.417	100	26 417.308	100	22 702.036

Geometry of volume
ABBREVIATIONS

b = breadth of base
h = perpendicular height
l = length of base
r = length of radius

$\pi = 3.1416$

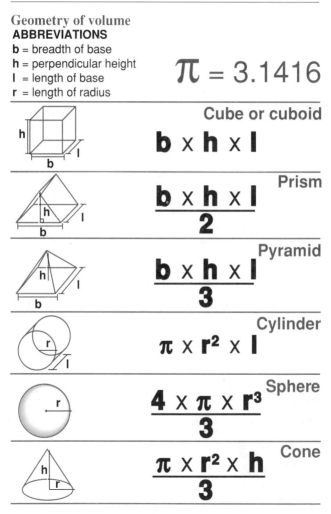

Cube or cuboid

$$b \times h \times l$$

Prism

$$\frac{b \times h \times l}{2}$$

Pyramid

$$\frac{b \times h \times l}{3}$$

Cylinder

$$\pi \times r^2 \times l$$

Sphere

$$\frac{4 \times \pi \times r^3}{3}$$

Cone

$$\frac{\pi \times r^2 \times h}{3}$$

4: Weight

Formulas

Below are listed the multiplication/division factors for converting units of length from imperial to metric, and vice versa, and from one unit to another in the same system. Note that two kinds of factors are given: quick, for an approximate conversion that can be made without a calculator; and accurate, for an exact conversion. The term "weight" differs in everyday use from its scientific use. In everyday terms, we use weight to describe how much substance an object has. In science, the term "mass" is used to describe this quantity of matter. Weight is used to describe the gravitational force on an object and is equal to its mass multiplied by the gravitational field strength. In scientific terms, mass remains constant but weight varies according to the strength of gravity. All units that follow are strictly units of mass rather than weight, apart from the pressure units kg/cm^2 and PSI.

		Quick	**Accurate**
	Grams (g) Grains (gr)		
	g ⟶ gr	× 15	× 15.432
	gr ⟶ g	÷ 15	× 0.065
	Ounces (oz) Grams (g)		
	oz ⟶ g	× 28	× 28.349
	g ⟶ oz	÷ 28	× 0.035

			Quick	**Accurate**
	Ounces troy (oz tr)			
	Grams (g)			
	oz tr ⟶ g		× 31	× 31.103
	g ⟶ oz tr		÷ 31	× 0.032
	Stones (st)			
	Kilograms (kg)			
	st ⟶ kg		× 6	× 6.350
	kg ⟶ st		÷ 6	× 0.157
	Long (UK) tons (l t)			
	Tonnes (t)			
	l t ⟶ t		× 1	× 1.016
	t ⟶ l t		÷ 1	× 0.984
	Kilograms (kg)			
	Pounds (lb)			
	kg ⟶ lb		× 2	× 2.205
	lb ⟶ kg		÷ 2	× 0.454
	Kilograms per square centimetre (kg/cm^2)			
	Pounds per square inch (PSI)			
	kg/cm^2 ⟶ PSI		× 14	× 14.223
	PSI ⟶ kg/cm^2		÷ 14	× 0.070
	Tonnes (t)			
	Short (US) tons (sh t)			
	t ⟶ sh t		× 1	× 1.102
	sh t ⟶ t		÷ 1	× 0.907
	Ounces troy (oz tr)			
	Ounces (oz)			
	oz tr ⟶ oz		× 1	× 1.097
	oz ⟶ oz tr		÷ 1	× 0.911

Conversion tables

The tables below can be used to convert units of weight from one measuring system to another. The units included in the tables are UK imperial, US imperial troy, and metric.

Grains to Grams		Ounces to Grams		Ounces troy to Grams	
gr	g	oz	g	oz tr	g
1	0.065	1	28.349	1	31.103
2	0.130	2	56.699	2	62.207
3	0.194	3	85.048	3	93.310
4	0.259	4	113.398	4	124.414
5	0.324	5	141.747	5	155.517
6	0.389	6	170.097	6	186.621
7	0.454	7	198.446	7	217.724
8	0.518	8	226.796	8	248.829
9	0.583	9	255.145	9	279.931
10	0.648	10	283.495	10	311.035
20	1.296	20	566.990	20	622.070
30	1.944	30	850.485	30	933.104
40	2.592	40	1133.980	40	1244.139
50	3.240	50	1417.475	50	1555.174
60	3.888	60	1700.970	60	1866.209
70	4.536	70	1984.465	70	2177.243
80	5.184	80	2267.960	80	2488.278
90	5.832	90	2551.455	90	2799.313
100	6.480	100	2834.900	100	3110.348

Pounds to Kilograms	
lb	kg
1	0.454
2	0.907
3	1.361
4	1.814
5	2.268
6	2.722
7	3.175
8	3.629
9	4.082
10	4.536
20	9.072
30	13.608
40	18.144
50	22.680
60	27.216
70	31.751
80	36.287
90	40.823
100	45.359

Pounds per square inch to Kilograms per square centimetre	
PSI	kg/cm^2
10	0.703
15	1.055
20	1.406
22	1.547
24	1.687
26	1.828
28	1.986
30	2.109
32	2.250
34	2.390
36	2.531
38	2.671
40	2.812
45	3.164
50	3.515

Stones to Kilograms	
st	kg
1	6.350
2	12.700
3	19.050
4	25.401
5	31.751
6	38.101
7	44.452
8	50.802
9	57.152
10	63.502
20	127.006
30	190.509
40	254.012
50	317.515
60	381.018
70	444.521
80	508.023
90	571.526
100	635.029

Imperial and metric units of weight (continued)

Short (US) tons to Tonnes		Long (UK) tons to Tonnes		Grams to Grains	
sh t	t	l t	t	g	gr
1	0.907	1	1.016	1	15.432
2	1.814	2	2.032	2	30.865
3	2.721	3	3.048	3	46.297
4	3.628	4	4.064	4	61.729
5	4.535	5	5.080	5	77.162
6	5.443	6	6.096	6	92.594
7	6.350	7	7.112	7	108.027
8	7.257	8	8.128	8	123.459
9	8.164	9	9.144	9	138.891
10	9.071	10	10.160	10	154.324
20	18.143	20	20.320	20	308.647
30	27.215	30	30.481	30	462.971
40	36.287	40	40.641	40	617.294
50	45.359	50	50.802	50	771.618
60	54.431	60	60.962	60	925.942
70	63.502	70	71.123	70	1080.265
80	72.574	80	81.283	80	1234.589
90	81.646	90	91.444	90	1388.912
100	90.718	100	101.604	100	1543.236

Grams to Ounces		Grams to Ounces troy		Kilograms to Pounds	
g	oz	g	oz tr	kg	lb
1	0.035	1	0.032	1	2.205
2	0.071	2	0.064	2	4.409
3	0.106	3	0.096	3	6.614
4	0.141	4	0.129	4	8.818
5	0.176	5	0.161	5	11.023
6	0.212	6	0.193	6	13.228
7	0.247	7	0.225	7	15.432
8	0.282	8	0.257	8	17.637
9	0.317	9	0.289	9	19.842
10	0.353	10	0.322	10	22.046
20	0.705	20	0.643	20	44.092
30	1.058	30	0.965	30	66.139
40	1.411	40	1.286	40	88.185
50	1.764	50	1.608	50	110.231
60	2.116	60	1.929	60	132.277
70	2.469	70	2.251	70	154.324
80	2.822	80	2.572	80	176.370
90	3.175	90	2.894	90	198.416
100	3.527	100	3.215	100	220.462

Imperial and metric units of weight (continued)

Kilograms per square centimetre to Pounds per square inch	
kg/cm^2	PSI
0.6	8.534
0.8	11.378
1.0	14.223
1.2	17.068
1.4	19.912
1.6	22.757
1.8	25.601
2.0	28.446
2.2	31.291
2.4	34.135
2.6	36.980
2.8	39.824
3.0	42.669
3.2	45.514
3.5	49.781

Kilograms to Stones	
kg	st
1	0.157
2	0.315
3	0.472
4	0.630
5	0.787
6	0.945
7	1.102
8	1.260
9	1.417
10	1.574
20	3.149
30	4.724
40	6.299
50	7.874
60	9.448
70	11.023
80	12.598
90	14.173
100	15.747

Tonnes to Short (US) tons	
t	sh t
1	1.102
2	2.205
3	3.307
4	4.409
5	5.512
6	6.614
7	7.716
8	8.818
9	9.921
10	11.023
20	22.046
30	33.069
40	44.092
50	55.116
60	66.139
70	77.162
80	88.185
90	99.208
100	110.231

Tonnes to Long (UK) tons	
t	l t
1	0.984
2	1.968
3	2.953
4	3.937
5	4.921
6	5.905
7	6.889
8	7.874
9	8.858
10	9.842
20	19.684
30	29.526
40	39.368
50	49.211
60	59.052
70	68.894
80	78.737
90	88.579
100	98.421

Ounces to Ounces troy	
oz	oz tr
1	0.911
2	1.823
3	2.734
4	3.646
5	4.557
6	5.468
7	6.380
8	7.291
9	8.203
10	9.114
20	18.229
30	27.344
40	36.458
50	45.573
60	54.687
70	63.802
80	72.917
90	82.031
100	91.146

Ounces troy to Ounces	
oz tr	oz
1	1.097
2	2.194
3	3.291
4	4.389
5	5.486
6	6.583
7	7.680
8	8.777
9	9.874
10	10.971
20	21.943
30	32.914
40	43.886
50	54.857
60	65.828
70	76.800
80	87.771
90	98.743
100	109.714

Periodic table

The periodic table is a means of classifying and comparing chemical elements. Substances as different as hydrogen, calcium and gold are all elements; each has distinctive properties and cannot be split chemically into a simpler form.

The table groups elements into seven rows or periods. Elements in the vertical columns, or groups, have similar properties. For example, the first element in any period (called an alkali metal) is reactive; while the last element (a noble, or inert, gas) is almost totally non-reactive.

1 H								
3 Li	4 Be							
11 Na	12 Mg							
19 K	20 Ca	21 Sc	22 Ti	23 V	24 Cr	25 Mn	26 Fe	27 Co
37 Rb	38 Sr	39 Y	40 Zr	41 Nb	42 Mo	43 Tc	44 Ru	45 Rh
55 Cs	56 Ba	57-71 -	72 Hf	73 Ta	74 W	75 Re	76 Os	77 Ir
87 Fr	88 Ra	89-103 -	104 Unq	105 Unp	106 Unh	107 Uns	108 Uno	109 Une

57 La	58 Ce	59 Pr	60 Nd	61 Pm	62 Sm	63 Eu
89 Ac	90 Th	91 Pa	92 U	93 Np	94 Pu	95 Am

The elements are listed in the table in order of their atomic numbers, from 1 to 109 (appearing in the upper left-hand corner of each box). The atomic number represents the number of protons the element has in its nucleus.

The two bottom rows are the lanthanides (57–71) and the actinides (89–103). These are separate because they have such similar properties that they fit the space of only two elements in the main table.

																	2 He
												5 B	6 C	7 N	8 O	9 F	10 Ne
												13 Al	14 Si	15 P	16 S	17 Cl	18 Ar
28 Ni	29 Cu	30 Zn	31 Ga	32 Ge	33 As	34 Se	35 Br	36 Kr
46 Pd	47 Ag	48 Cd	49 In	50 Sn	51 Sb	52 Te	53 I	54 Xe
78 Pt	79 Au	80 Hg	81 Tl	82 Pb	83 Bi	84 Po	85 At	86 Rn

64 Gd	65 Tb	66 Dy	67 Ho	68 Er	69 Tm	70 Yb	71 Lu
96 Cm	97 Bk	98 Cf	99 Es	100 Fm	101 Md	102 No	103 Lr

Chemical elements

On the following pages, the elements are listed in three separate ways: **1** by atomic number; **2** by element name; and **3** by letter symbol. Each listing includes the atomic number, element name, symbol, and atomic weight (or relative atomic mass) of each element.

* Indicates atomic weight of the isotope with the longest known half-life.

1 BY ATOMIC NUMBER

Atomic No.	Name	Symbol	Atomic weight
1	Hydrogen	H	1.007 9
2	Helium	He	4.002 6
3	Lithium	Li	6.941
4	Beryllium	Be	9.012 18
5	Boron	B	10.81
6	Carbon	C	12.011
7	Nitrogen	N	14.006 7
8	Oxygen	O	15.999 4
9	Fluorine	F	18.998 4
10	Neon	Ne	20.179
11	Sodium	Na	22.989 77
12	Magnesium	Mg	24.305
13	Aluminium	Al	26.981 54
14	Silicon	Si	28.085 5
15	Phosphorus	P	30.973 76
16	Sulphur	S	32.064
17	Chlorine	Cl	35.453
18	Argon	Ar	39.948
19	Potassium	K	39.098 3
20	Calcium	Ca	40.08

Atomic No.	Name	Symbol	Atomic weight
21	Scandium	Sc	44.955 9
22	Titanium	Ti	47.9
23	Vanadium	V	50.941 4
24	Chromium	Cr	51.996
25	Manganese	Mn	54.938
26	Iron	Fe	55.847
27	Cobalt	Co	58.933 2
28	Nickel	Ni	58.71
29	Copper	Cu	63.546
30	Zinc	Zn	65.381
31	Gallium	Ga	69.72
32	Germanium	Ge	72.59
33	Arsenic	As	74.921 6
34	Selenium	Se	78.96
35	Bromine	Br	79.904
36	Krypton	Kr	83.8
37	Rubidium	Rb	85.467 8
38	Strontium	Sr	87.62
39	Yttrium	Y	88.905 9
40	Zirconium	Zr	91.22
41	Niobium	Nb	92.906 4
42	Molybdenum	Mo	95.94
43	Technetium	Tc	96.906 4*
44	Ruthenium	Ru	101.07
45	Rhodium	Rh	102.905 5
46	Palladium	Pd	106.4
47	Silver	Ag	107.868
48	Cadmium	Cd	112.41
49	Indium	In	114.82

Atomic No.	Name	Symbol	Atomic weight
50	Tin	Sn	118.69
51	Antimony	Sb	121.75
52	Tellurium	Te	127.6
53	Iodine	I	126.904 5
54	Xenon	Xe	131.3
55	Caesium	Cs	132.905 4
56	Barium	Ba	137.33
57	Lanthanum	La	138.905 5
58	Caerium	Ce	140.12
59	Praseodymium	Pr	140.907 7
60	Neodymium	Nd	144.24
61	Promethium	Pm	144.912 8*
62	Samarium	Sm	150.35
63	Europium	Eu	151.96
64	Galolinium	Gd	157.25
65	Terbium	Tb	158.925 4
66	Dysprosium	Dy	162.5
67	Holmium	Ho	164.930 4
68	Erbium	Er	167.26
69	Thulium	Tm	168.934 2
70	Ytterbium	Yb	173.04
71	Lutetium	Lu	174.97
72	Hafnium	Hf	178.49
73	Tantalum	Ta	180.947 9
74	Tungsten	W	183.85
75	Rhenium	Re	186.207
76	Osmium	Os	190.2
77	Iridium	Ir	192.22
78	Platinum	Pt	195.09

Atomic No.	Name	Symbol	Atomic weight
79	Gold	Au	196.966 5
80	Mercury	Hg	200.59
81	Thallium	Tl	204.37
82	Lead	Pb	207.19
83	Bismuth	Bi	208.980 4
84	Polonium	Po	208.982 4*
85	Astatine	At	209.987 0*
86	Radon	Rn	222.017 6*
87	Francium	Fr	223.019 7*
88	Radium	Ra	226.025 4*
89	Actinium	Ac	227.027 8*
90	Thorium	Th	232.038 1
91	Protoactinium	Pa	231.035 9
92	Uranium	U	238.029*
93	Neptunium	Np	237.048 2*
94	Plutonium	Pu	244.064 2*
95	Americium	Am	243.061 4*
96	Curium	Cm	247.070 3*
97	Berkelium	Bk	247.070 3*
98	Californium	Cf	251.079 6*
99	Einsteinium	Es	254.088 0*
100	Fermium	Fm	257.095 1*
101	Mendelevium	Md	258.099*
102	Nobelium	No	259.101*
103	Lawrencium	Lr	260.105*
104	Unnilquadium	Unq	261.109*
105	Unnilpentium	Unp	262.114*
106	Unnilhexium	Unh	263.120*
107	Unnilseptium	Uns	262*

Atomic No.	Name	Symbol	Atomic weight
108	Unniloctium	Uno	265
109	Unnilennium	Une	266*

2 BY ELEMENT NAME

Name	Atomic No.	Symbol	Atomic weight
Actinium	89	Ac	227.027 8*
Aluminium	13	Al	26.981 54
Americium	95	Am	243.061 4*
Antimony	51	Sb	121.75
Argon	18	Ar	39.948
Arsenic	33	As	74.921 6
Astatine	85	At	209.987 0*
Barium	56	Ba	137.33
Berkelium	97	Bk	247.070 3*
Beryllium	4	Be	9.012 18
Bismuth	83	Bi	208.980 4
Boron	5	B	10.81
Bromine	35	Br	79.904
Cadmium	48	Cd	112.41
Caesium	55	Cs	132.905 4
Calcium	20	Ca	40.08
Californium	98	Cf	251.079 6*
Carbon	6	C	12.011
Cerium	58	Ce	140.12
Chlorine	17	Cl	35.453
Chromium	24	Cr	51.996
Cobalt	27	Co	58.933 2

Name	Atomic No.	Symbol	Atomic weight
Copper	29	Cu	63.546
Curium	96	Cm	247.703*
Dysprosium	66	Dy	162.5
Einsteinium	99	Es	254.088 0*
Erbium	68	Er	167.26
Europium	63	Eu	151.96
Fermium	100	Fm	257.095 1*
Fluorine	9	F	18.998 4
Francium	87	Fr	223.019 7*
Gallium	31	Ga	69.72
Galolinium	64	Gd	157.25
Germanium	32	Ge	72.59
Gold	79	Au	196.966 5
Hafnium	72	Hf	178.49
Helium	2	He	4.002 6
Holmium	67	Ho	164.930 4
Hydrogen	1	H	1.007 9
Iodine	53	I	126.904 5
Indium	49	In	114.82
Iridium	77	Ir	192.22
Iron	26	Fe	55.847
Krypton	36	Kr	83.8
Lanthanum	57	La	138.905 5
Lawrencium	103	Lr	260.105*
Lead	82	Pb	207.19
Lithium	3	Li	6.941
Lutetium	71	Lu	174.97
Magnesium	12	Mg	24.305

Name	Atomic No.	Symbol	Atomic weight
Manganese	25	Mn	54.938
Mendelevium	101	Md	258.099*
Mercury	80	Hg	200.59
Molybdenum	42	Mo	95.94
Neodymium	60	Nd	144.24
Neon	10	Ne	20.179
Neptunium	93	Np	237.048 2*
Nickel	28	Ni	58.71
Niobium	41	Nb	92.906 4
Nitrogen	7	N	14.006 7
Nobelium	102	No	259.101*
Oxygen	8	O	15.999 4
Osmium	76	Os	190.2
Palladium	46	Pd	106.4
Phosphorus	15	P	30.973 76
Platinum	78	Pt	195.09
Plutonium	94	Pu	244.064 2*
Polonium	84	Po	208.982 4*
Potassium	19	K	39.098 3
Praseodymium	59	Pr	140.907 7
Promethium	61	Pm	144.912 8*
Protoactinium	91	Pa	231.035 9
Radium	88	Ra	226.0254*
Radon	86	Rn	222.017 6*
Rhenium	75	Re	186.207
Rhodium	45	Rh	102.905 5
Rubidium	37	Rb	85.467 8
Ruthenium	44	Ru	101.07

Name	Atomic No.	Symbol	Atomic weight
Samarium	62	Sm	150.35
Scandium	21	Sc	44.955 9
Selenium	34	Se	78.96
Sodium	11	Na	22.989 77
Silicon	14	Si	28.085 5
Silver	47	Ag	107.868
Strontium	38	Sr	87.62
Sulphur	16	S	32.064
Tantalum	73	Ta	180.947 9
Technetium	43	Tc	96.906 4*
Tellurium	52	Te	127.6
Terbium	65	Tb	158.925 4
Thallium	81	Tl	204.37
Thorium	90	Th	232.038 1
Thulium	69	Tm	168.934 2
Tin	50	Sn	118.69
Titanium	22	Ti	47.9
Tungsten	74	W	183.85
Unnilhexium	106	Unh	263.120*
Unnilennium	109	Une	266*
Unniloctium	108	Uno	265
Unnilquadium	104	Unq	261.109*
Unnilseptium	107	Uns	262*
Unnilpentium	105	Unp	262.114*
Uranium	92	U	238.029*
Vanadium	23	V	50.941 4
Xenon	54	Xe	131.3
Ytterbium	70	Yb	173.04

Name	Atomic No.	Symbol	Atomic weight
Yttrium	39	Y	88.905 9
Zinc	30	Zn	65.381
Zirconium	40	Zr	91.22

3 BY LETTER SYMBOL

Symbol	Atomic No.	Name	Atomic weight
Ac	89	Actinium	227.027 8*
Ag	47	Silver	107.868
Al	13	Aluminium	26.981 54
Am	95	Americium	243.061 4*
Ar	18	Argon	39.948
As	33	Arsenic	74.921 6
At	85	Astatine	209.987 0*
Au	79	Gold	196.966 5
B	5	Boron	10.81
Ba	56	Barium	137.33
Be	4	Beryllium	9.012 18
Bk	97	Berkelium	247.070 3*
Bi	83	Bismuth	208.980 4
Br	35	Bromine	79.904
C	6	Carbon	12.011
Ca	20	Calcium	40.08
Cd	48	Cadmium	112.41
Ce	58	Cerium	140.12
Cf	98	Californium	251.079 6*
Cl	17	Chlorine	35.453
Cm	96	Curium	247.070 3*
Co	27	Cobalt	58.933 2

Symbol	Atomic No.	Name	Atomic weight
Cs	55	Caesium	132.905 4
Cu	29	Copper	63.546
Dy	66	Dysprosium	162.5
Er	68	Erbium	167.26
Es	99	Einsteinium	254.088*
Eu	63	Europium	151.96
F	9	Fluorine	18.998 4
Fe	26	Iron	55.847
Fm	100	Fermium	257.095 1*
Fr	87	Francium	223.019 7*
Ga	31	Gallium	69.72
Gd	64	Galolinium	157.25
Ge	32	Germanium	72.59
H	1	Hydrogen	1.007 9
He	2	Helium	4.002 6
Hf	72	Hafnium	178.49
Hg	80	Mercury	200.59
Ho	67	Holmium	164.930 4
I	53	Iodine	126.904 5
In	49	Indium	114.82
Ir	77	Iridium	192.22
K	19	Potassium	39.098 3
Kr	36	Krypton	83.8
La	57	Lanthanum	138.905 5
Li	3	Lithium	6.941
Lr	103	Lawrencium	260.105*
Lu	71	Lutetium	174.97
Md	101	Mendelevium	258.099*
Mg	12	Magnesium	24.305

Symbol	Atomic No.	Name	Atomic weight
Mn	25	Manganese	54.938
Mo	42	Molybdenum	95.94
N	7	Nitrogen	14.006 7
Na	11	Sodium	22.989 77
Nb	41	Niobium	92.906 4
Nd	60	Neodymium	144.24
Ne	10	Neon	20.179
Ni	28	Nickel	58.71
No	102	Nobelium	259.101*
Np	93	Neptunium	237.048 2*
O	8	Oxygen	15.999 4
Os	76	Osmium	190.2
P	15	Phosphorus	30.973 76
Pa	91	Protoactinium	231.035 9
Pb	82	Lead	207.19
Pd	46	Palladium	106.4
Pm	61	Promethium	144.912 8*
Po	84	Polonium	208.982 4*
Pr	59	Praseodymium	140.907 7
Pt	78	Platinum	195.09
Pu	94	Plutonium	244.064 2*
Ra	88	Radium	226.025 4*
Rb	37	Rubidium	85.467 8
Re	75	Rhenium	186.207
Rh	45	Rhodium	102.905 5
Rn	86	Radon	222.017 6*
Ru	44	Ruthenium	101.07
S	16	Sulphur	32.064
Sb	51	Antimony	121.75

Symbol	Atomic No.	Name	Atomic weight
Sc	21	Scandium	44.955 9
Se	34	Selenium	78.96
Si	14	Silicon	28.085 5
Sm	62	Samarium	150.35
Sn	50	Tin	118.69
Sr	38	Strontium	87.62
Ta	73	Tantalum	180.947 9
Tb	65	Terbium	158.925 4
Tc	43	Technetium	96.906 4*
Te	52	Tellurium	127.6
Th	90	Thorium	232.038 1
Ti	22	Titanium	47.9
Tl	81	Thallium	204.37
Tm	69	Thulium	168.934 2
U	92	Uranium	238.029*
Une	109	Unnilennium	266*
Unh	106	Unnilhexium	263.120*
Uno	108	Unniloctium	no data
Unp	105	Unnilpentium	262.114*
Unq	104	Unnilquadium	261.109*
Uns	107	Unnilseptium	262*
V	23	Vanadium	50.941 4
W	74	Tungsten	183.85
Xe	54	Xenon	131.3
Y	39	Yttrium	88.905 9
Yb	70	Ytterbium	173.04
Zn	30	Zinc	65.381
Zr	40	Zirconium	91.22

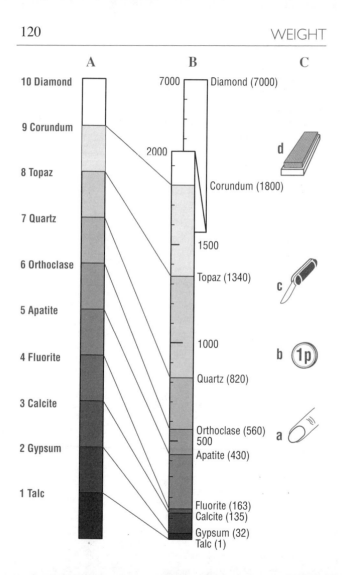

Scales of hardness

Solids vary in their degree of hardness, which indicates their resistance to being scratched or cut.

A Mohs' scale

Mohs' scale is used to measure the relative hardness of minerals. The framework uses the 10 minerals – talc to diamond – shown in the scale. Each of these minerals is assigned a numerical value from 1 to 10: the higher the number, the harder the mineral.

Order is determined by the ability of a mineral to scratch all those that have a lower number and to be scratched by those with a higher number. Once this is established, it is possible to place all other minerals on the scale by means of the same scratching procedure.

B Knoop scale

Another system of measuring the hardness of minerals is the Knoop scale. The Knoop scale gives absolute rather than relative measurements. Readings on this scale are made by measuring the size of the indentation made by a diamond-shaped device dropped on the material.

Again, the higher the number the harder the substance, but the intervals between minerals and levels of hardness differ greatly from scale to scale. Minerals with values between 1 and 7 on Mohs' scale fall below 1000 on the Knoop scale, and between 8 and 9 fall below 2000, but diamond falls at 7000.

C Common-object scale

A simple way of measuring hardness uses common objects, whose hardness on the Mohs' scale is known:

a) fingernail (2–2.5 Mohs')   c) knife blade (5–6)

b) penny (4)   d) knife sharpener (8–9)

5: Energy

Formulas
Below are listed the multiplication/division factors for
converting units of energy from imperial to metric, and
vice versa. Note that two kinds of factors are given:
quick, for an approximate conversion that can be made
without a calculator; and accurate, for an exact
conversion.

			Quick	**Accurate**
Kilowatts (kW) Horsepower (hp)				
	kW → hp		× 1.5	× 1.341
	hp → kW		÷ 1.5	× 0.746
Calories (cal) Joules (J)				
	cal → J		× 4	× 4.187
	J → cal		÷ 4	× 0.239
Kilocalories (kcal) Kilojoules (kJ)				
	kcal → kJ		× 4	× 4.187
	kJ → kcal		÷ 4	× 0.239

Conversion tables

The tables below can be used to convert units of energy
from metric to imperial systems, and vice versa.

Horsepower to Kilowatts		Kilowatts to Horsepower	
hp	kW	kW	hp
1	0.746	1	1.341
2	1.491	2	2.682
3	2.237	3	4.023
4	2.983	4	5.364
5	3.729	5	6.705
6	4.474	6	8.046
7	5.220	7	9.387
8	5.966	8	10.728
9	6.711	9	12.069
10	7.457	10	13.410
20	14.914	20	26.820
30	22.371	30	40.231
40	29.828	40	53.641
50	37.285	50	67.051
60	44.742	60	80.461
70	52.199	70	93.871
80	59.656	80	107.280
90	67.113	90	120.690
100	74.570	100	134.100

Metric units of energy

J	cal
Joules to Calories international	
1	0.239
2	0.478
3	0.716
4	0.955
5	1.194
6	1.433
7	1.672
8	1.911
9	2.150
10	2.388
20	4.777
30	7.165
40	9.554
50	11.942
60	14.330
70	16.719
80	19.108
90	21.496
100	23.885

kJ	kcal
Kilojoules to Kilocalories international	
1	0.239
2	0.478
3	0.716
4	0.955
5	1.194
6	1.433
7	1.672
8	1.911
9	2.150
10	2.388
20	4.777
30	7.165
40	9.554
50	11.942
60	14.330
70	16.719
80	19.108
90	21.496
100	23.885

Calories international to Joules	
cal	J
1	4.187
2	8.374
3	12.560
4	16.747
5	20.934
6	25.121
7	29.308
8	33.494
9	37.681
10	41.868
20	83.736
30	125.604
40	167.472
50	209.340
60	251.208
70	293.076
80	334.944
90	376.812
100	418.680

Kilocalories international to Kilojoules	
kcal	kJ
1	4.187
2	8.374
3	12.560
4	16.747
5	20.934
6	25.121
7	29.308
8	33.494
9	37.681
10	41.868
20	83.736
30	125.604
40	167.472
50	209.340
60	251.208
70	293.076
80	334.944
90	376.812
100	418.680

Electromagnetic spectrum

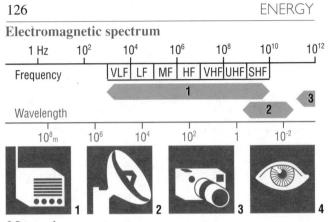

Measuring energy

Light and radio waves, X-rays, and other forms of energy are transmitted through space as electromagnetic waves. These waves have alternating high and low points – crests and troughs – like actual waves. The distance between wave crests is called the wavelength; this is measured in metres. Frequency refers to the number of waves per second passing a certain point; this is measured in hertz (Hz).

Above is an electromagnetic spectrum, showing the different forms of energy in order of frequency and wavelength. The top part of the diagram shows the frequency in hertz; the lower part measures the wavelength in metres.

1 Radio waves

These waves transmit television and radio signals. This section of the spectrum is divided into bands, from VLF (very low frequency) – used for time signals – to SHF (super-high frequency) – used for space and satellite communication.

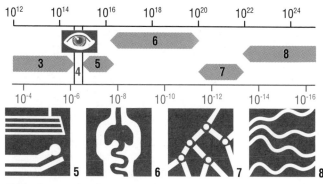

2 Radar and microwaves
Radar bounces waves off objects, allowing unseen objects to be seen; microwaves can cook food quickly.
3 Infrared waves
These waves are emitted by all hot objects.
4 Visible light
The band of visible light from red to violet.
5 Ultraviolet light
In small amounts, these waves produce vitamin D and cause skin to tan; in larger amounts they can damage living cells.
6 X-rays
Used to photograph the internal structures of the body.
7 Gamma rays
Emitted during the decay of some radioisotopes, these waves can be very damaging to the body.
8 Cosmic rays
Caused by nuclear explosions and reactions in space, nearly all of these waves are absorbed by the Earth's atmosphere.

Earthquakes

Earthquake magnitude is measured in units on the
Richter scale, which measures the amount of energy
released. Each year there are more than 300 000 earth
tremors with Richter magnitudes of 2 to 2.9. An
earthquake of 8.5 or higher occurs about every 5 to 10
years.

Intensity

The intensity of an earthquake is measured on the
Mercalli scale; the numbers refer to an earthquake's
effect at a specific place on the Earth's surface.

Below are listed numbers on the Mercalli scale and
the characteristics of each.

No.	Characteristic
I	instrumental (detected only by seismograph)
II	feeble (noticed only by people at rest)
III	slight (similar to vibrations from a passing truck)
IV	moderate (felt indoors, parked cars rock)
V	rather strong (felt generally, waking sleepers)
VI	strong (trees sway, some damage)
VII	very strong (general alarm, walls crack)
VIII	destructive (walls collapse)
IX	ruinous (some houses collapse, ground cracks)
X	disastrous (buildings destroyed, rails bend)
XI	very disastrous (landslides, few buildings survive)
XII	catastrophic (total destruction)

Listed below are the Mercalli and Richter scales, with equivalents in joules, and a table comparing the Richter scale with joules.

Mercalli	Richter	Joules	Richter	Joules
I	<3.5	$<1.6 \times 10^7$ J	0	6.3×10^{-2} J
II	3.5	1.6×10^7 J	1	1.6×10 J
III	4.2	7.5×10^8 J	2	4.0×10^3 J
IV	4.5	4.0×10^9 J	3	1.0×10^6 J
V	4.8	2.1×10^{10} J	4	2.5×10^8 J
VI	5.4	5.7×10^{11} J	5	6.3×10^{10} J
VII	6.1	2.8×10^{13} J	6	1.6×10^{13} J
VIII	6.5	2.5×10^{14} J	7	4.0×10^{15} J
IX	6.9	2.3×10^{15} J	8	1.0×10^{18} J
X	7.3	2.1×10^{16} J	9	2.5×10^{20} J
XI	8.1	1.7×10^{18} J	10	6.3×10^{22} J
XII	>8.1	$>1.7 \times 10^{18}$ J		

Actual earthquakes
The table below lists the year of selected earthquakes in different parts of the world and where they occurred, as well as the Richter magnitude of each.

Earthquakes	Richter
Assam, India (1897)	8.7
Alaska, USA (1964)	8.6
Concepción, Chile (1960)	8.5
San Francisco, USA (1906)	8.25
Mexico City, Mexico (1985)	8.1
Guatemala (1976)	7.9
Tangshan, China (1976)	7.6
Messina, Italy (1908)	7.5
Vrancea, Romania (1977)	7.2
San Francisco, USA (1989)	6.9

Decibels

The loudness of a sound is measured by the size of its vibrations; this is measured in decibels (dB).

Decibel scale

The dB scale is relative and increases exponentially, beginning with the smallest sound change that can be heard by humans (0–1 dB). A 20 dB sound is 10 times louder than a 10 dB sound; a 30 dB sound is 100 times as loud as that. Noises at the level of 120–130 dB can cause pain in humans; higher levels can cause permanent ear damage. The dB ratings (at certain distances) of some common noises are listed on page 135.

Wave amplitude

Amplitude is the distance between a wave peak or trough and an intermediate line of equilibrium (**a**). The greater the amount of energy transmitted in a sound wave, the greater is the wave's amplitude and the louder the sound heard.

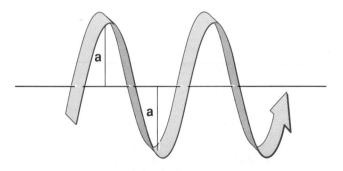

A 0 dB   human minimum audibility

B 30 dB   soft whisper at 5 m

C 50 dB   inside urban home

D 55 dB   light traffic at 15 m

E 60 dB   conversation at 1 m

F 85 dB   pneumatic drill at 15 m

G 90 dB   heavy traffic at 15 m

H 100 dB   loud shout at 15 m

I 105 dB   aeroplane take-off at 600 m

J 117 dB   inside full-volume disco

K 120 dB   aeroplane take-off at 60 m

L 130 dB   pain threshold for humans

M 140 dB   aeroplane take-off at 30 m

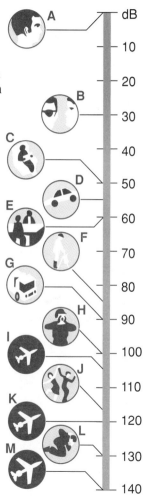

Energy needs by activity

	Activity	☐ Women	■ Men
A	Sleeping	1230 kJ; 55 kcal	1272 kJ; 65 kcal
B	Sitting	1293 kJ; 70 kcal	1377 kJ; 90 kcal
C	Standing	1419 kJ; 100 kcal	1502 kJ; 120 kcal
D	Walking	1754 kJ; 180 kcal	1921 kJ; 220 kcal
E	Walking (uphill)	1507 kJ; 360 kcal	1842 kJ; 440 kcal
F	Running	1759 kJ; 420 kcal	2512 kJ; 600 kcal

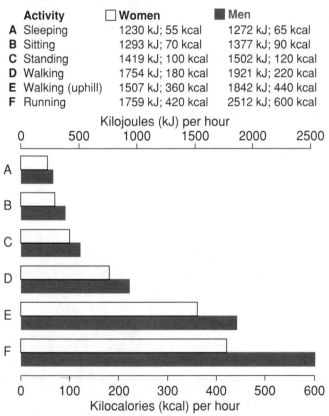

Men use more kilocalories than women for all activities because men have more weight to carry around, and because women usually have more body fat and so need less energy to retain body heat.

Energy values of selected foods

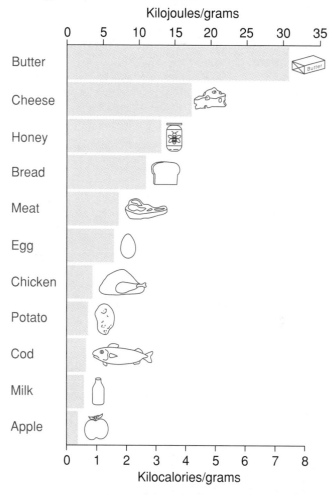

6: Temperature

Systems of measurement
Below, the different systems of temperature
measurement are compared: Fahrenheit (°F), Celsius
(°C), Réaumur (°r), Rankine (°R) and Kelvin (K). Also
listed are the formulas for converting temperature
measurements from one system to another.

Formulas

°F ➡ °C	(°F−32)÷1.8	°r ➡ K	(°r×1.25)+273.16
°C ➡ °F	(°C×1.8)+32	°R ➡ K	°R÷1.8
°F ➡ K	(°F+459.67)÷1.8	K ➡ °F	(K×1.8)−459.67
°C ➡ K	°C+273.16	K ➡ °C	K−273.16

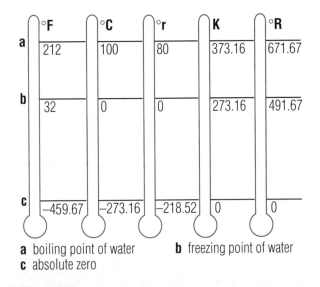

	°F	°C	°r	K	°R
a	212	100	80	373.16	671.67
b	32	0	0	273.16	491.67
c	−459.67	−273.16	−218.52	0	0

a boiling point of water   **b** freezing point of water
c absolute zero

Conversion tables
The tables below list the equivalent units of temperature in the Fahrenheit, Celsius and Kelvin systems.

Fahrenheit to Celsius to Kelvin			Fahrenheit to Celsius to Kelvin		
°F	°C	K	°F	°C	K
−40.0	−40	233	−4.0	−20	253
−38.2	−39	234	−2.2	−19	254
−36.4	−38	235	−0.4	−18	255
−34.6	−37	236	1.4	−17	256
−32.8	−36	237	3.2	−16	257
−31.0	−35	238	5.0	−15	258
−29.2	−34	239	6.8	−14	259
−27.4	−33	240	8.6	−13	260
−25.6	−32	241	10.4	−12	261
−23.8	−31	242	12.2	−11	262
−22.0	−30	243	14.0	−10	263
−20.2	−29	244	15.8	−9	264
−18.4	−28	245	17.6	−8	265
−16.6	−27	246	19.4	−7	266
−14.8	−26	247	21.2	−6	267
−13.0	−25	248	23.0	−5	268
−11.2	−24	249	24.8	−4	269
−9.4	−23	250	26.6	−3	270
−7.6	−22	251	28.4	−2	271
−5.8	−21	252	30.2	−1	272

Fahrenheit, Celsius and Kelvin unit equivalents (continued)

Fahrenheit to Celsius to Kelvin		
°F	°C	K
32.0	0	273
33.8	1	274
35.6	2	275
37.4	3	276
39.2	4	277
41.0	5	278
42.8	6	279
44.6	7	280
46.4	8	281
48.2	9	282
50.0	10	283
51.8	11	284
53.6	12	285
55.4	13	286
57.2	14	287
59.0	15	288
60.8	16	289
62.6	17	290
64.4	18	291
66.2	19	292

Fahrenheit to Celsius to Kelvin		
°F	°C	K
68.0	20	293
69.8	21	294
71.6	22	295
73.4	23	296
75.2	24	297
77.0	25	298
78.8	26	299
80.6	27	300
82.4	28	301
84.2	29	302
86.0	30	303
87.8	31	304
89.6	32	305
91.4	33	306
93.2	34	307
95.0	35	308
96.8	36	309
98.6	37	310
100.4	38	311
102.2	39	312

Fahrenheit to Celsius to Kelvin			Fahrenheit to Celsius to Kelvin		
°F	°C	K	°F	°C	K
104.0	40	313	140.0	60	333
105.8	41	314	141.8	61	334
107.6	42	315	143.6	62	335
109.4	43	316	145.4	63	336
111.2	44	317	147.2	64	337
113.0	45	318	149.0	65	338
114.8	46	319	150.8	66	339
116.6	47	320	152.6	67	340
118.4	48	321	154.4	68	341
120.2	49	322	156.2	69	342
122.0	50	323	158.0	70	343
123.8	51	324	159.8	71	344
125.6	55	325	161.6	72	345
127.4	53	326	163.4	73	346
129.2	54	327	165.2	74	347
131.0	55	328	167.0	75	348
132.8	56	329	168.8	76	349
134.6	57	330	170.6	77	350
136.4	58	331	172.4	78	351
138.2	59	332	174.2	79	352

Fahrenheit, Celsius and Kelvin unit equivalents (continued)

Fahrenheit to Celsius to Kelvin				Fahrenheit to Celsius to Kelvin		
°F	°C	K		°F	°C	K
176.0	80	353		212.0	100	373
177.8	81	354		213.8	101	374
179.6	82	355		215.6	102	375
181.4	83	356		217.4	103	376
183.2	84	357		219.2	104	377
185.0	85	358		221.0	105	378
186.8	86	359		222.8	106	379
188.6	87	360		224.6	107	380
190.4	88	361		226.4	108	381
192.2	89	362		228.2	109	382
194.0	90	363		230.0	110	383
195.8	91	364		231.8	111	384
197.6	92	365		233.6	112	385
199.4	93	366		235.4	113	386
201.2	94	367		237.2	114	387
203.0	95	368		239.0	115	388
204.8	96	369		240.8	116	389
206.6	97	370		242.6	117	390
208.4	98	371		244.4	118	391
210.2	99	372		246.2	119	392

Useful temperatures

Quick temperature reference

Condition	°C	°F
Water freezes	0	32
Mild winter day	10	50
Warm spring day	20	68
Hot summer day	30	86
Body temperature	37	98.6
Heat wave	40	104
Water boils	100	212

Boiling points

Boiling point is the temperature at which a liquid bubbles and becomes gas. The temperature at which things boil depends on their molucular structure and on atmospheric pressure, which is determined by altitude. The higher the altitude, the lower the boiling point. The boiling point of water varies in different parts of the world.

Place	Altitude	Water boils at
London, UK	Sea level	100° C
Dead sea	-1296 ft	101° C
Denver, Col	5280 ft	95° C
Quito, Ecuador	9350 ft	90° C
Lhasa, Tibet	12,087 ft	87° C
Mt Everest	29,002 ft	71° C

7: Time

Units of time
Listed below are the names of time periods that are artificially derived, as opposed to astronomical periods.

Time periods			
Name	**Period**	**Name**	**Period**
bicentennial	200 years	olympiad	4 years
biennial	2 years	quadrennial	4 years
century	100 years	quadricentennial	400 years
day	24 hours	quincentennial	500 years
decade	10 years	quindecennial	15 years
centennial	100 years	quinquennial	5 years
decennial	10 years	semicentennial	50 years
duodecennial	12 years	septennial	7 years
half-century	50 years	sesquicentennial	150 years
half-decade	5 years	sexennial	6 years
half-millennium	500 years	tricennial	30 years
hour	60 minutes	triennial	3 years
leap year	366 days	undecennial	11 years
millennium	1,000 years	vicennial	20 years
minute	60 seconds	week	7 days
month	28-31 days	year	365 days
novennial	9 years	year	12 months
octennial	8 years	year	52 weeks

Days, hours, minutes
Below are listed the basic subdivisions of a day and their equivalents.

1 day = 24 hours = 1440 minutes = 86 400 seconds

1 hour = $1/24$ day = 60 minutes = 3600 seconds
1 minute = $1/1440$ day = $1/60$ hour = 60 seconds
1 second = $1/86\,400$ day = $1/3600$ hour = $1/60$ minute

Time intervals

Names for recurring time intervals	
annual	yearly
biannual	twice a year (at unequally spaced intervals)
bimonthly	every two months: twice a month
biweekly	every two weeks: twice a week
diurnal	daily: of a day
perennial	occurring year after year
semiannual	every six months (at equally spaced intervals)
semidiurnal	twice a day
semiweekly	twice a week
trimonthly	every three months
triweekly	every three weeks: three times a week
thrice weekly	three times a week

Seconds

Greater precision in measuring time has required seconds (s) to be broken down into smaller units, using standard metric prefixes.

1 tetrasecond (Ts)	10^{12} s	31 689 years
1 gigasecond (Gs)	10^9 s	31.7 years
1 megasecond (Ms)	10^6 s	11.6 days
1 kilosecond (ks)	10^3 s	16.67 minutes
1 millisecond (ms)	10^{-3} s	0.001 seconds
1 microsecond (μs)	10^{-6} s	0.000 001 seconds

1 nanosecond (ns)	10^{-9} s	0.000 000 001
1 picosecond (ps)	10^{-12} s	0.000 000 000 001
1 femtosecond (fs)	10^{-15} s	0.000 000 000 000 001
1 attosecond (as)	10^{-18} s	0.000 000 000 000 000 001

Astronomical time

Time can be measured by motion; in fact, the motion of the Earth, Sun, Moon and stars provided humans with the first means of measuring time.

Years, months, days

Sidereal times are calculated by the Earth's position according to fixed stars. The anomalistic year is measured according to the Earth's orbit in relation to the perihelion (Earth's minimum distance to the Sun). Tropical times refer to the apparent passage of the Sun and the actual passage of the Moon across the Earth's equatorial plane. The synodic month is based on the phases of the Moon. Solar time (as in a mean solar day) refers to periods of darkness and light averaged over a year.

Time	Days	Hours	Minutes	Seconds
sidereal year	365	6	9	10
anomalistic year	365	6	13	53
tropical year	365	5	48	45
sidereal month	27	7	43	11
tropical month	27	7	43	5
synodic month	29	12	44	3
mean solar day	0	24	0	0
sidereal day	0	23	56	4

Equinox and solstice

The inclination of the Earth to its plane of rotation around the Sun produces variations in the lengths of day and night at different times of the year. Solstices are when the Sun appears to be overhead at midday at the maximum distances north and south of the Equator. At the summer solstice, days are longest and nights are shortest; this is reversed at the winter solstice. Equinoxes are when day and night are equal everywhere; at these times, the Sun appears overhead at midday at the Equator.

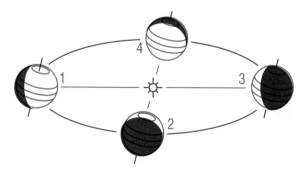

The table lists the dates of the solstices and equinox in each hemisphere, keyed by number to the diagram below, which shows the Earth at four points in its orbit.

	Date	Northern	Southern
1	21 June	summer solstice	winter solstice
2	23 Sept.	autumn equinox	spring equinox
3	22 Dec.	winter solstice	summer solstice
4	21 March	spring equinox	autumn equinox

Years and seasons

Seasonal variations are another result of the inclination
of the Earth's axis to its plane of rotation around the
Sun. Parts of the globe tilted away from the Sun
receive less radiant energy per unit area than those
receiving rays more directly. The table below lists the
seasonal equivalents in the two hemispheres, keyed by
number to the diagram on the previous page.

Northern	**Southern**
1 summer	winter
2 autumn	spring
3 winter	summer
4 spring	autumn

Length of days

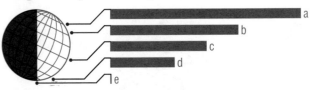

a Arctic Circle: 66° 33′N – 24 hours daylight
b 49° 3′N – 16 hours daylight
c The Equator: 0° – 12 hours daylight
d 49° 3′S – 8 hours daylight
e Antarctic Circle: 66° 33′S – 0 hours daylight

The diagram above illustrates the variety in the length
of the day (21 June) at different latitudes (° = degrees;
′ = minutes). On this day, the northern hemisphere
receives the maximum hours of daylight; the southern
hemisphere, the minimum.

Types of calendar
The number of days in a year varies among cultures and from year to year.

Gregorian
The Gregorian calendar is a 16th-century adaptation of the Julian calendar devised in the 1st century BC. The year in this calendar is based on the solar year, which lasts about 365 $^1/_4$ days. In this system, years whose number is not divisible by 4 have 365 days, as do centennial years unless the figures before the noughts are divisible by 4. All other years have 366 days; these are leap years.

Below are the names of the months and number of days for a non-leap year.

January	31	July	31
February	28*	August	31
March	31	September	30
April	30	October	31
May	31	November	30
June	30	December	31

* 29 in leap years.

Jewish
A year in the Jewish calendar has 13 months if its number, when divided by 9, leaves 0, 3, 6, 8, 11, 14 or 17; otherwise, it has 12 months. The year is based on the lunar year, but its number of months varies to keep broadly in line with the solar cycle. Its precise number of days is fixed with reference to particular festivals that must not fall on certain days of the week.

Below are the names of the months and number of days in each for the year 5471, a 12-month year (1980 AD in Gregorian).

Tishri	30	Nisan	30
Cheshvan	29*	Iyar	29
Kislev	29*	Sivan	30
Tevet	29	Tammuz	29
Shevat	30	Av	30
Adar	29†	Elul	29

* 30 in some years.
† In 13-month years, the month Veadar, with 29 days, falls between Adar and Nisan.

Muslim

A year in the Muslim calendar has 355 days if its number, when divided by 30, leaves 2, 5, 7, 10, 13, 16, 18, 21, 24, 26 or 29; otherwise it has 354 days. As in the Jewish calendar, years are based on the lunar cycle.

Below are the names of the months and number of days in each for the Muslim year 1401 (1980 AD in Gregorian).

Muharram	30	Rajab	30
Safar	29	Sha'ban	29
Rabi'I	30	Ramadan	30
Rabi'II	29	Shawwal	29
Jumada I	30	Dhu l-Qa'dah	30
Jumada II	29	Dhu l-Hijja	30*

* 29 in some years.

Perpetual calendar

How to use the calendar To discover on which day of the week any date between the years 1780 and 2014 falls, look up the year in the key and the letter shown to the right will indicate which of the calendars A–N you should consult.

Key:

Year		Year		Year	
1780	N	1805	C	1830	F
1781	B	1806	D	1831	G
1782	C	1807	E	1832	H
1783	D	1808	M	1833	C
1784	L	1809	A	1834	D
1785	G	1810	B	1835	E
1786	A	1811	C	1836	M
1787	B	1812	K	1837	A
1788	J	1813	F	1838	B
1789	E	1814	G	1839	C
1790	F	1815	A	1840	K
1791	G	1816	I	1841	F
1792	H	1817	D	1842	G
1793	C	1818	E	1843	A
1794	D	1819	F	1844	I
1795	E	1820	N	1845	D
1796	M	1821	B	1846	E
1797	A	1822	C	1847	F
1798	B	1823	D	1848	N
1799	C	1824	L	1849	B
1800	D	1825	G	1850	C
1801	E	1826	A	1851	D
1802	F	1827	B	1852	L
1803	G	1828	J	1853	G
1804	H	1829	E	1854	A

1855	B	1887	G	1919	D
1856	J	1888	H	1920	L
1857	E	1889	C	1921	G
1858	F	1890	D	1922	A
1859	G	1891	E	1923	B
1860	H	1892	M	1924	J
1861	C	1893	A	1925	E
1862	D	1894	B	1926	F
1863	E	1895	C	1927	G
1864	M	1896	K	1928	H
1865	A	1897	F	1929	C
1866	B	1898	G	1930	D
1867	C	1899	A	1931	E
1868	K	1900	B	1932	M
1869	F	1901	C	1933	A
1870	G	1902	D	1934	B
1871	A	1903	E	1935	C
1872	I	1904	M	1936	K
1873	D	1905	A	1937	F
1874	E	1906	B	1938	G
1875	F	1907	C	1939	A
1876	N	1908	K	1940	I
1877	B	1909	F	1941	D
1878	C	1910	G	1942	E
1879	D	1911	A	1943	F
1880	L	1912	I	1944	N
1881	G	1913	D	1945	B
1882	A	1914	E	1946	C
1883	B	1915	F	1947	D
1884	J	1916	N	1948	L
1885	E	1917	B	1949	G
1886	F	1918	C	1950	A

Year		Year		Year	
1951	B	1983	G	2015	E
1952	J	1984	H	2016	M
1953	E	1985	C	2017	A
1954	F	1986	D	2018	B
1955	G	1987	E	2019	C
1956	H	1988	M	2020	K
1957	C	1989	A	2021	F
1958	D	1990	B	2022	G
1959	E	1991	C	2023	A
1960	M	1992	K	2024	I
1961	A	1993	F	2025	D
1962	B	1994	G	2026	E
1963	C	1995	A	2027	F
1964	K	1996	I	2028	N
1965	F	1997	D	2029	B
1966	G	1998	E	2030	C
1967	A	1999	F	2031	D
1968	I	2000	N	2032	L
1969	D	2001	B	2033	G
1970	E	2002	C	2034	A
1971	F	2003	D	2035	B
1972	N	2004	L	2036	J
1973	B	2005	G	2037	E
1974	C	2006	A	2038	F
1975	D	2007	B	2039	G
1976	L	2008	J	2040	H
1977	G	2009	E	2041	C
1978	A	2010	F	2042	D
1979	B	2011	G	2043	E
1980	J	2012	H	2044	M
1981	E	2013	C	2045	A
1982	F	2014	D	2046	B

A 1786 1797 1809 1815 1826 1837 1843
       1854 1865 1871 1882 1893 1899 1905

JANUARY
S	M	T	W	T	F	S
1	2	3	4	5	6	7
8	9	10	11	12	13	14
15	16	17	18	19	20	21
22	23	24	25	26	27	28
29	30	31				

FEBRUARY
S	M	T	W	T	F	S
				1	2	3	4
5	6	7	8	9	10	11
12	13	14	15	16	17	18
19	20	21	22	23	24	25
26	27	28				

MARCH
S	M	T	W	T	F	S
			1	2	3	4
5	6	7	8	9	10	11
12	13	14	15	16	17	18
19	20	21	22	23	24	25
26	27	28	29	30	31	

APRIL
S	M	T	W	T	F	S
						1
2	3	4	5	6	7	8
9	10	11	12	13	14	15
16	17	18	19	20	21	22
23	24	25	26	27	28	29
30						

MAY
S	M	T	W	T	F	S
	1	2	3	4	5	6
7	8	9	10	11	12	13
14	15	16	17	18	19	20
21	22	23	24	25	26	27
28	29	30	31			

JUNE
S	M	T	W	T	F	S
				1	2	3
4	5	6	7	8	9	10
11	12	13	14	15	16	17
18	19	20	21	22	23	24
25	26	27	28	29	30	

**1911 1922 1933 1939 1950 1961 1967
1978 1989 1995 2006 2017 2023 2034
2045**

A

JULY
S	M	T	W	T	F	S
						1
2	3	4	5	6	7	8
9	10	11	12	13	14	15
16	17	18	19	20	21	22
23	24	25	26	27	28	29
30	31					

AUGUST
S	M	T	W	T	F	S
		1	2	3	4	5
6	7	8	9	10	11	12
13	14	15	16	17	18	19
20	21	22	23	24	25	26
27	28	29	30	31		

SEPTEMBER
S	M	T	W	T	F	S
					1	2
3	4	5	6	7	8	9
10	11	12	13	14	15	16
17	18	19	20	21	22	23
24	25	26	27	28	29	30

OCTOBER
S	M	T	W	T	F	S
1	2	3	4	5	6	7
8	9	10	11	12	13	14
15	16	17	18	19	20	21
22	23	24	25	26	27	28
29	30	31				

NOVEMBER
S	M	T	W	T	F	S
			1	2	3	4
5	6	7	8	9	10	11
12	13	14	15	16	17	18
19	20	21	22	23	24	25
26	27	28	29	30		

DECEMBER
S	M	T	W	T	F	S
					1	2
3	4	5	6	7	8	9
10	11	12	13	14	15	16
17	18	19	20	21	22	23
24	25	26	27	28	29	30
31						

B 1781 1787 1798 1810 1821 1827 1838
       1849 1855 1866 1877 1883 1894 1900

JANUARY
S	M	T	W	T	F	S
	1	2	3	4	5	6
7	8	9	10	11	12	13
14	15	16	17	18	19	20
21	22	23	24	25	26	27
28	29	30	31			

FEBRUARY
S	M	T	W	T	F	S
				1	2	3
4	5	6	7	8	9	10
11	12	13	14	15	16	17
18	19	20	21	22	23	24
25	26	27	28			

MARCH
S	M	T	W	T	F	S
				1	2	3
4	5	6	7	8	9	10
11	12	13	14	15	16	17
18	19	20	21	22	23	24
25	26	27	28	29	30	31

APRIL
S	M	T	W	T	F	S
1	2	3	4	5	6	7
8	9	10	11	12	13	14
15	16	17	18	19	20	21
22	23	24	25	26	27	28
29	30					

MAY
S	M	T	W	T	F	S
		1	2	3	4	5
6	7	8	9	10	11	12
13	14	15	16	17	18	19
20	21	22	23	24	25	26
27	28	29	30	31		

JUNE
S	M	T	W	T	F	S
					1	2
3	4	5	6	7	8	9
10	11	12	13	14	15	16
17	18	19	20	21	22	23
24	25	26	27	28	29	30

1906 1917 1923 1934 1945 1951 1962 1973 1979 1990 2001 2007 2018 2029 2035 2046

B

JULY

S	M	T	W	T	F	S
1	2	3	4	5	6	7
8	9	10	11	12	13	14
15	16	17	18	19	20	21
22	23	24	25	26	27	28
29	30	31				

AUGUST

S	M	T	W	T	F	S
				1	2	3	4
5	6	7	8	9	10	11
12	13	14	15	16	17	18
19	20	21	22	23	24	25
26	27	28	29	30	31	

SEPTEMBER

S	M	T	W	T	F	S
						1
2	3	4	5	6	7	8
9	10	11	12	13	14	15
16	17	18	19	20	21	22
23	24	25	26	27	28	29
30						

OCTOBER

S	M	T	W	T	F	S
	1	2	3	4	5	6
7	8	9	10	11	12	13
14	15	16	17	18	19	20
21	22	23	24	25	26	27
28	29	30	31			

NOVEMBER

S	M	T	W	T	F	S
				1	2	3
4	5	6	7	8	9	10
11	12	13	14	15	16	17
18	19	20	21	22	23	24
25	26	27	28	29	30	

DECEMBER

S	M	T	W	T	F	S
						1
2	3	4	5	6	7	8
9	10	11	12	13	14	15
16	17	18	19	20	21	22
23	24	25	26	27	28	29
30	31					

C 1782 1793 1799 1805 1811 1822 1833
    1839 1850 1861 1867 1878 1889 1895

JANUARY
S	M	T	W	T	F	S
		1	2	3	4	5
6	7	8	9	10	11	12
13	14	15	16	17	18	19
20	21	22	23	24	25	26
27	28	29	30	31		

FEBRUARY
S	M	T	W	T	F	S
					1	2
3	4	5	6	7	8	9
10	11	12	13	14	15	16
17	18	19	20	21	22	23
24	25	26	27	28		

MARCH
S	M	T	W	T	F	S
					1	2
3	4	5	6	7	8	9
10	11	12	13	14	15	16
17	18	19	20	21	22	23
24	25	26	27	28	29	30
31						

APRIL
S	M	T	W	T	F	S
	1	2	3	4	5	6
7	8	9	10	11	12	13
14	15	16	17	18	19	20
21	22	23	24	25	26	27
28	29	30				

MAY
S	M	T	W	T	F	S
			1	2	3	4
5	6	7	8	9	10	11
12	13	14	15	16	17	18
19	20	21	22	23	24	25
26	27	28	29	30	31	

JUNE
S	M	T	W	T	F	S
						1
2	3	4	5	6	7	8
9	10	11	12	13	14	15
16	17	18	19	20	21	22
23	24	25	26	27	28	29
30						

1901 1907 1918 1929 1935 1946 1957
1963 1974 1985 1991 2002 2013 2019
2030 2041

C

JULY
S	M	T	W	T	F	S
	1	2	3	4	5	6
7	8	9	10	11	12	13
14	15	16	17	18	19	20
21	22	23	24	25	26	27
28	29	30	31			

AUGUST
S	M	T	W	T	F	S
				1	2	3
4	5	6	7	8	9	10
11	12	13	14	15	16	17
18	19	20	21	22	23	24
25	26	27	28	29	30	31

SEPTEMBER
S	M	T	W	T	F	S
1	2	3	4	5	6	7
8	9	10	11	12	13	14
15	16	17	18	19	20	21
22	23	24	25	26	27	28
29	30					

OCTOBER
S	M	T	W	T	F	S
		1	2	3	4	5
6	7	8	9	10	11	12
13	14	15	16	17	18	19
20	21	22	23	24	25	26
27	28	29	30	31		

NOVEMBER
S	M	T	W	T	F	S
					1	2
3	4	5	6	7	8	9
10	11	12	13	14	15	16
17	18	19	20	21	22	23
24	25	26	27	28	29	30

DECEMBER
S	M	T	W	T	F	S
1	2	3	4	5	6	7
8	9	10	11	12	13	14
15	16	17	18	19	20	21
22	23	24	25	26	27	28
29	30	31				

D

| 1783 | 1794 | 1800 | 1806 | 1817 | 1823 | 1834 |
| 1845 | 1851 | 1862 | 1873 | 1879 | 1890 | 1902 |

JANUARY

S	M	T	W	T	F	S
			1	2	3	4
5	6	7	8	9	10	11
12	13	14	15	16	17	18
19	20	21	22	23	24	25
26	27	28	29	30	31	

FEBRUARY

S	M	T	W	T	F	S
						1
2	3	4	5	6	7	8
9	10	11	12	13	14	15
16	17	18	19	20	21	22
23	24	25	26	27	28	

MARCH

S	M	T	W	T	F	S
						1
2	3	4	5	6	7	8
9	10	11	12	13	14	15
16	17	18	19	20	21	22
23	24	25	26	27	28	29
30	31					

APRIL

S	M	T	W	T	F	S
		1	2	3	4	5
6	7	8	9	10	11	12
13	14	15	16	17	18	19
20	21	22	23	24	25	26
27	28	29	30			

MAY

S	M	T	W	T	F	S
				1	2	3
4	5	6	7	8	9	10
11	12	13	14	15	16	17
18	19	20	21	22	23	24
25	26	27	28	29	30	31

JUNE

S	M	T	W	T	F	S
1	2	3	4	5	6	7
8	9	10	11	12	13	14
15	16	17	18	19	20	21
22	23	24	25	26	27	28
29	30					

1913 1919 1930 1941 1947 1958 1969 1975 1986 1997 2003 2014 2025 2031 2042

D

JULY

S	M	T	W	T	F	S
		1	2	3	4	5
6	7	8	9	10	11	12
13	14	15	16	17	18	19
20	21	22	23	24	25	26
27	28	29	30	31		

AUGUST

S	M	T	W	T	F	S
					1	2
3	4	5	6	7	8	9
10	11	12	13	14	15	16
17	18	19	20	21	22	23
24	25	26	27	28	29	30
31						

SEPTEMBER

S	M	T	W	T	F	S
	1	2	3	4	5	6
7	8	9	10	11	12	13
14	15	16	17	18	19	20
21	22	23	24	25	26	27
28	29	30				

OCTOBER

S	M	T	W	T	F	S
			1	2	3	4
5	6	7	8	9	10	11
12	13	14	15	16	17	18
19	20	21	22	23	24	25
26	27	28	29	30	31	

NOVEMBER

S	M	T	W	T	F	S
						1
2	3	4	5	6	7	8
9	10	11	12	13	14	15
16	17	18	19	20	21	22
23	24	25	26	27	28	29
30						

DECEMBER

S	M	T	W	T	F	S
	1	2	3	4	5	6
7	8	9	10	11	12	13
14	15	16	17	18	19	20
21	22	23	24	25	26	27
28	29	30	31			

E
1789	1795	1801	1807	1818	1829	1835
1846	1857	1863	1874	1885	1891	1903

JANUARY

S	M	T	W	T	F	S
				1	2	3
4	5	6	7	8	9	10
11	12	13	14	15	16	17
18	19	20	21	22	23	24
25	26	27	28	29	30	31

FEBRUARY

S	M	T	W	T	F	S
1	2	3	4	5	6	7
8	9	10	11	12	13	14
15	16	17	18	19	20	21
22	23	24	25	26	27	28

MARCH

S	M	T	W	T	F	S
1	2	3	4	5	6	7
8	9	10	11	12	13	14
15	16	17	18	19	20	21
22	23	24	25	26	27	28
29	30	31				

APRIL

S	M	T	W	T	F	S
			1	2	3	4
5	6	7	8	9	10	11
12	13	14	15	16	17	18
19	20	21	22	23	24	25
26	27	28	29	30		

MAY

S	M	T	W	T	F	S
					1	2
3	4	5	6	7	8	9
10	11	12	13	14	15	16
17	18	19	20	21	22	23
24	25	26	27	28	29	30
31						

JUNE

S	M	T	W	T	F	S
	1	2	3	4	5	6
7	8	9	10	11	12	13
14	15	16	17	18	19	20
21	22	23	24	25	26	27
28	29	30				

1914 1925 1931 1942 1953 1959 1970
1981 1987 1998 2009 2015 2026 2037
2043

E

JULY

S	M	T	W	T	F	S
		1	2	3	4	
5	6	7	8	9	10	11
12	13	14	15	16	17	18
19	20	21	22	23	24	25
26	27	28	29	30	31	

AUGUST

S	M	T	W	T	F	S
						1
2	3	4	5	6	7	8
9	10	11	12	13	14	15
16	17	18	19	20	21	22
23	24	25	26	27	28	29
30	31					

SEPTEMBER

S	M	T	W	T	F	S
		1	2	3	4	5
6	7	8	9	10	11	12
13	14	15	16	17	18	19
20	21	22	23	24	25	26
27	28	29	30			

OCTOBER

S	M	T	W	T	F	S
				1	2	3
4	5	6	7	8	9	10
11	12	13	14	15	16	17
18	19	20	21	22	23	24
25	26	27	28	29	30	31

NOVEMBER

S	M	T	W	T	F	S
1	2	3	4	5	6	7
8	9	10	11	12	13	14
15	16	17	18	19	20	21
22	23	24	25	26	27	28
29	30					

DECEMBER

S	M	T	W	T	F	S
		1	2	3	4	5
6	7	8	9	10	11	12
13	14	15	16	17	18	19
20	21	22	23	24	25	26
27	28	29	30	31		

F 1790 1802 1813 1819 1830 1841 1847
1858 1869 1875 1886 1897 1909 1915

JANUARY

S	M	T	W	T	F	S
					1	2
3	4	5	6	7	8	9
10	11	12	13	14	15	16
17	18	19	20	21	22	23
24	25	26	27	28	29	30
31						

FEBRUARY

S	M	T	W	T	F	S
	1	2	3	4	5	6
7	8	9	10	11	12	13
14	15	16	17	18	19	20
21	22	23	24	25	26	27
28						

MARCH

S	M	T	W	T	F	S
	1	2	3	4	5	6
7	8	9	10	11	12	13
14	15	16	17	18	19	20
21	22	23	24	25	26	27
28	29	30	31			

APRIL

S	M	T	W	T	F	S
				1	2	3
4	5	6	7	8	9	10
11	12	13	14	15	16	17
18	19	20	21	22	23	24
25	26	27	28	29	30	

MAY

S	M	T	W	T	F	S
						1
2	3	4	5	6	7	8
9	10	11	12	13	14	15
16	17	18	19	20	21	22
23	24	25	26	27	28	29
30	31					

JUNE

S	M	T	W	T	F	S
		1	2	3	4	5
6	7	8	9	10	11	12
13	14	15	16	17	18	19
20	21	22	23	24	25	26
27	28	29	30			

1926 1937 1943 1954 1965 1971 1982
1993 1999 2010 2021 2027 2038 **F**

JULY

S	M	T	W	T	F	S
				1	2	3
4	5	6	7	8	9	10
11	12	13	14	15	16	17
18	19	20	21	22	23	24
25	26	27	28	29	30	31

AUGUST

S	M	T	W	T	F	S
1	2	3	4	5	6	7
8	9	10	11	12	13	14
15	16	17	18	19	20	21
22	23	24	25	26	27	28
29	30	31				

SEPTEMBER

S	M	T	W	T	F	S
			1	2	3	4
5	6	7	8	9	10	11
12	13	14	15	16	17	18
19	20	21	22	23	24	25
26	27	28	29	30		

OCTOBER

S	M	T	W	T	F	S
					1	2
3	4	5	6	7	8	9
10	11	12	13	14	15	16
17	18	19	20	21	22	23
24	25	26	27	28	29	30
31						

NOVEMBER

S	M	T	W	T	F	S
	1	2	3	4	5	6
7	8	9	10	11	12	13
14	15	16	17	18	19	20
21	22	23	24	25	26	27
28	29	30				

DECEMBER

S	M	T	W	T	F	S
			1	2	3	4
5	6	7	8	9	10	11
12	13	14	15	16	17	18
19	20	21	22	23	24	25
26	27	28	29	30	31	

G 1785 1791 1803 1814 1825 1831 1842
1853 1859 1870 1881 1887 1898 1910

JANUARY

S	M	T	W	T	F	S
						1
2	3	4	5	6	7	8
9	10	11	12	13	14	15
16	17	18	19	20	21	22
23	24	25	26	27	28	29
30	31					

FEBRUARY

S	M	T	W	T	F	S
		1	2	3	4	5
6	7	8	9	10	11	12
13	14	15	16	17	18	19
20	21	22	23	24	25	26
27	28					

MARCH

S	M	T	W	T	F	S
		1	2	3	4	5
6	7	8	9	10	11	12
13	14	15	16	17	18	19
20	21	22	23	24	25	26
27	28	29	30	31		

APRIL

S	M	T	W	T	F	S
					1	2
3	4	5	6	7	8	9
10	11	12	13	14	15	16
17	18	19	20	21	22	23
24	25	26	27	28	29	30

MAY

S	M	T	W	T	F	S
1	2	3	4	5	6	7
8	9	10	11	12	13	14
15	16	17	18	19	20	21
22	23	24	25	26	27	28
29	30	31				

JUNE

S	M	T	W	T	F	S
			1	2	3	4
5	6	7	8	9	10	11
12	13	14	15	16	17	18
19	20	21	22	23	24	25
26	27	28	29	30		

1921 1927 1938 1949 1955 1966 1977
1983 1994 2005 2011 2022 2033 2039 G

JULY

S	M	T	W	T	F	S
					1	2
3	4	5	6	7	8	9
10	11	12	13	14	15	16
17	18	19	20	21	22	23
24	25	26	27	28	29	30
31						

AUGUST

S	M	T	W	T	F	S
	1	2	3	4	5	6
7	8	9	10	11	12	13
14	15	16	17	18	19	20
21	22	23	24	25	26	27
28	29	30	31			

SEPTEMBER

S	M	T	W	T	F	S
				1	2	3
4	5	6	7	8	9	10
11	12	13	14	15	16	17
18	19	20	21	22	23	24
25	26	27	28	29	30	

OCTOBER

S	M	T	W	T	F	S
						1
2	3	4	5	6	7	8
9	10	11	12	13	14	15
16	17	18	19	20	21	22
23	24	25	26	27	28	29
30	31					

NOVEMBER

S	M	T	W	T	F	S
		1	2	3	4	5
6	7	8	9	10	11	12
13	14	15	16	17	18	19
20	21	22	23	24	25	26
27	28	29	30			

DECEMBER

S	M	T	W	T	F	S
				1	2	3
4	5	6	7	8	9	10
11	12	13	14	15	16	17
18	19	20	21	22	23	24
25	26	27	28	29	30	31

H

1792 1804 1832 1860 1888
1928 1956 1984 2012 2040

JANUARY

S	M	T	W	T	F	S
1	2	3	4	5	6	7
8	9	10	11	12	13	14
15	16	17	18	19	20	21
22	23	24	25	26	27	28
29	30	31				

FEBRUARY

S	M	T	W	T	F	S
			1	2	3	4
5	6	7	8	9	10	11
12	13	14	15	16	17	18
19	20	21	22	23	24	25
26	27	28	29			

MARCH

S	M	T	W	T	F	S
				1	2	3
4	5	6	7	8	9	10
11	12	13	14	15	16	17
18	19	20	21	22	23	24
25	26	27	28	29	30	31

APRIL

S	M	T	W	T	F	S
1	2	3	4	5	6	7
8	9	10	11	12	13	14
15	16	17	18	19	20	21
22	23	24	25	26	27	28
29	30					

MAY

S	M	T	W	T	F	S
		1	2	3	4	5
6	7	8	9	10	11	12
13	14	15	16	17	18	19
20	21	22	23	24	25	26
27	28	29	30	31		

JUNE

S	M	T	W	T	F	S
					1	2
3	4	5	6	7	8	9
10	11	12	13	14	15	16
17	18	19	20	21	22	23
24	25	26	27	28	29	30

H

JULY
S	M	T	W	T	F	S
1	2	3	4	5	6	7
8	9	10	11	12	13	14
15	16	17	18	19	20	21
22	23	24	25	26	27	28
29	30	31				

AUGUST
S	M	T	W	T	F	S
			1	2	3	4
5	6	7	8	9	10	11
12	13	14	15	16	17	18
19	20	21	22	23	24	25
26	27	28	29	30	31	

SEPTEMBER
S	M	T	W	T	F	S
						1
2	3	4	5	6	7	8
9	10	11	12	13	14	15
16	17	18	19	20	21	22
23	24	25	26	27	28	29
30						

OCTOBER
S	M	T	W	T	F	S
	1	2	3	4	5	6
7	8	9	10	11	12	13
14	15	16	17	18	19	20
21	22	23	24	25	26	27
28	29	30	31			

NOVEMBER
S	M	T	W	T	F	S
				1	2	3
4	5	6	7	8	9	10
11	12	13	14	15	16	17
18	19	20	21	22	23	24
25	26	27	28	29	30	

DECEMBER
S	M	T	W	T	F	S
						1
2	3	4	5	6	7	8
9	10	11	12	13	14	15
16	17	18	19	20	21	22
23	24	25	26	27	28	29
30	31					

I 1816 1844 1872 1912
       1940 1968 1996 2024

JANUARY
S	M	T	W	T	F	S
	1	2	3	4	5	6
7	8	9	10	11	12	13
14	15	16	17	18	19	20
21	22	23	24	25	26	27
28	29	30	31			

FEBRUARY
S	M	T	W	T	F	S
				1	2	3
4	5	6	7	8	9	10
11	12	13	14	15	16	17
18	19	20	21	22	23	24
25	26	27	28	29		

MARCH
S	M	T	W	T	F	S
					1	2
3	4	5	6	7	8	9
10	11	12	13	14	15	16
17	18	19	20	21	22	23
24	25	26	27	28	29	30
31						

APRIL
S	M	T	W	T	F	S
	1	2	3	4	5	6
7	8	9	10	11	12	13
14	15	16	17	18	19	20
21	22	23	24	25	26	27
28	29	30				

MAY
S	M	T	W	T	F	S
			1	2	3	4
5	6	7	8	9	10	11
12	13	14	15	16	17	18
19	20	21	22	23	24	25
26	27	28	29	30	31	

JUNE
S	M	T	W	T	F	S
						1
2	3	4	5	6	7	8
9	10	11	12	13	14	15
16	17	18	19	20	21	22
23	24	25	26	27	28	29
30						

I

JULY

S	M	T	W	T	F	S
	1	2	3	4	5	6
7	8	9	10	11	12	13
14	15	16	17	18	19	20
21	22	23	24	25	26	27
28	29	30	31			

AUGUST

S	M	T	W	T	F	S
				1	2	3
4	5	6	7	8	9	10
11	12	13	14	15	16	17
18	19	20	21	22	23	24
25	26	27	28	29	30	31

SEPTEMBER

S	M	T	W	T	F	S
1	2	3	4	5	6	7
8	9	10	11	12	13	14
15	16	17	18	19	20	21
22	23	24	25	26	27	28
29	30					

OCTOBER

S	M	T	W	T	F	S
		1	2	3	4	5
6	7	8	9	10	11	12
13	14	15	16	17	18	19
20	21	22	23	24	25	26
27	28	29	30	31		

NOVEMBER

S	M	T	W	T	F	S
					1	2
3	4	5	6	7	8	9
10	11	12	13	14	15	16
17	18	19	20	21	22	23
24	25	26	27	28	29	30

DECEMBER

S	M	T	W	T	F	S
1	2	3	4	5	6	7
8	9	10	11	12	13	14
15	16	17	18	19	20	21
22	23	24	25	26	27	28
29	30	31				

J 1788 1828 1856 1884 1924
       1952 1980 2008 2036

JANUARY
S	M	T	W	T	F	S
		1	2	3	4	5
6	7	8	9	10	11	12
13	14	15	16	17	18	19
20	21	22	23	24	25	26
27	28	29	30	31		

FEBRUARY
S	M	T	W	T	F	S
					1	2
3	4	5	6	7	8	9
10	11	12	13	14	15	16
17	18	19	20	21	22	23
24	25	26	27	28	29	

MARCH
S	M	T	W	T	F	S
						1
2	3	4	5	6	7	8
9	10	11	12	13	14	15
16	17	18	19	20	21	22
23	24	25	26	27	28	29
30	31					

APRIL
S	M	T	W	T	F	S
		1	2	3	4	5
6	7	8	9	10	11	12
13	14	15	16	17	18	19
20	21	22	23	24	25	26
27	28	29	30			

MAY
S	M	T	W	T	F	S
				1	2	3
4	5	6	7	8	9	10
11	12	13	14	15	16	17
18	19	20	21	22	23	24
25	26	27	28	29	30	31

JUNE
S	M	T	W	T	F	S
1	2	3	4	5	6	7
8	9	10	11	12	13	14
15	16	17	18	19	20	21
22	23	24	25	26	27	28
29	30					

J

JULY

S	M	T	W	T	F	S
		1	2	3	4	5
6	7	8	9	10	11	12
13	14	15	16	17	18	19
20	21	22	23	24	25	26
27	28	29	30	31		

AUGUST

S	M	T	W	T	F	S
					1	2
3	4	5	6	7	8	9
10	11	12	13	14	15	16
17	18	19	20	21	22	23
24	25	26	27	28	29	30
31						

SEPTEMBER

S	M	T	W	T	F	S
	1	2	3	4	5	6
7	8	9	10	11	12	13
14	15	16	17	18	19	20
21	22	23	24	25	26	27
28	29	30				

OCTOBER

S	M	T	W	T	F	S
			1	2	3	4
5	6	7	8	9	10	11
12	13	14	15	16	17	18
19	20	21	22	23	24	25
26	27	28	29	30	31	

NOVEMBER

S	M	T	W	T	F	S
						1
2	3	4	5	6	7	8
9	10	11	12	13	14	15
16	17	18	19	20	21	22
23	24	25	26	27	28	29
30						

DECEMBER

S	M	T	W	T	F	S
	1	2	3	4	5	6
7	8	9	10	11	12	13
14	15	16	17	18	19	20
21	22	23	24	25	26	27
28	29	30	31			

K   1812 1840 1868 1896 1908
1936 1864 1992 2020

JANUARY
S	M	T	W	T	F	S
				1	2	3	4
5	6	7	8	9	10	11
12	13	14	15	16	17	18
19	20	21	22	23	24	25
26	27	28	29	30	31	

FEBRUARY
S	M	T	W	T	F	S
						1
2	3	4	5	6	7	8
9	10	11	12	13	14	15
16	17	18	19	20	21	22
23	24	25	26	27	28	29

MARCH
S	M	T	W	T	F	S
1	2	3	4	5	6	7
8	9	10	11	12	13	14
15	16	17	18	19	20	21
22	23	24	25	26	27	28
29	30	31				

APRIL
S	M	T	W	T	F	S
			1	2	3	4
5	6	7	8	9	10	11
12	13	14	15	16	17	18
19	20	21	22	23	24	25
26	27	28	29	30		

MAY
S	M	T	W	T	F	S
					1	2
3	4	5	6	7	8	9
10	11	12	13	14	15	16
17	18	19	20	21	22	23
24	25	26	27	28	29	30
31						

JUNE
S	M	T	W	T	F	S
	1	2	3	4	5	6
7	8	9	10	11	12	13
14	15	16	17	18	19	20
21	22	23	24	25	26	27
28	29	30				

JULY

S	M	T	W	T	F	S
			1	2	3	4
5	6	7	8	9	10	11
12	13	14	15	16	17	18
19	20	21	22	23	24	25
26	27	28	29	30	31	

AUGUST

S	M	T	W	T	F	S
						1
2	3	4	5	6	7	8
9	10	11	12	13	14	15
16	17	18	19	20	21	22
23	24	25	26	27	28	29
30	31					

SEPTEMBER

S	M	T	W	T	F	S
		1	2	3	4	5
6	7	8	9	10	11	12
13	14	15	16	17	18	19
20	21	22	23	24	25	26
27	28	29	30			

OCTOBER

S	M	T	W	T	F	S
				1	2	3
4	5	6	7	8	9	10
11	12	13	14	15	16	17
18	19	20	21	22	23	24
25	26	27	28	29	30	31

NOVEMBER

S	M	T	W	T	F	S
1	2	3	4	5	6	7
8	9	10	11	12	13	14
15	16	17	18	19	20	21
22	23	24	25	26	27	28
29	30					

DECEMBER

S	M	T	W	T	F	S
		1	2	3	4	5
6	7	8	9	10	11	12
13	14	15	16	17	18	19
20	21	22	23	24	25	26
27	28	29	30	31		

L 1784 1824 1852 1880 1920
       1948 1976 2004 2032

JANUARY
S	M	T	W	T	F	S
				1	2	3
4	5	6	7	8	9	10
11	12	13	14	15	16	17
18	19	20	21	22	23	24
25	26	27	28	29	30	31

FEBRUARY
S	M	T	W	T	F	S
1	2	3	4	5	6	7
8	9	10	11	12	13	14
15	16	17	18	19	20	21
22	23	24	25	26	27	28
29						

MARCH
S	M	T	W	T	F	S
	1	2	3	4	5	6
7	8	9	10	11	12	13
14	15	16	17	18	19	20
21	22	23	24	25	26	27
28	29	30	31			

APRIL
S	M	T	W	T	F	S
				1	2	3
4	5	6	7	8	9	10
11	12	13	14	15	16	17
18	19	20	21	22	23	24
25	26	27	28	29	30	

MAY
S	M	T	W	T	F	S
						1
2	3	4	5	6	7	8
9	10	11	12	13	14	15
16	17	18	19	20	21	22
23	24	25	26	27	28	29
30	31					

JUNE
S	M	T	W	T	F	S
		1	2	3	4	5
6	7	8	9	10	11	12
13	14	15	16	17	18	19
20	21	22	23	24	25	26
27	28	29	30			

L

JULY

S	M	T	W	T	F	S
				1	2	3
4	5	6	7	8	9	10
11	12	13	14	15	16	17
18	19	20	21	22	23	24
25	26	27	28	29	30	31

AUGUST

S	M	T	W	T	F	S
1	2	3	4	5	6	7
8	9	10	11	12	13	14
15	16	17	18	19	20	21
22	23	24	25	26	27	28
29	30	31				

SEPTEMBER

S	M	T	W	T	F	S
			1	2	3	4
5	6	7	8	9	10	11
12	13	14	15	16	17	18
19	20	21	22	23	24	25
26	27	28	29	30		

OCTOBER

S	M	T	W	T	F	S
					1	2
3	4	5	6	7	8	9
10	11	12	13	14	15	16
17	18	19	20	21	22	23
24	25	26	27	28	29	30
31						

NOVEMBER

S	M	T	W	T	F	S
	1	2	3	4	5	6
7	8	9	10	11	12	13
14	15	16	17	18	19	20
21	22	23	24	25	26	27
28	29	30				

DECEMBER

S	M	T	W	T	F	S
			1	2	3	4
5	6	7	8	9	10	11
12	13	14	15	16	17	18
19	20	21	22	23	24	25
26	27	28	29	30	31	

M 1796 1808 1836 1864 1892 1904
1932 1960 1988 2016 2044

JANUARY
S	M	T	W	T	F	S
					1	2
3	4	5	6	7	8	9
10	11	12	13	14	15	16
17	18	19	20	21	22	23
24	25	26	27	28	29	30
31						

FEBRUARY
S	M	T	W	T	F	S
	1	2	3	4	5	6
7	8	9	10	11	12	13
14	15	16	17	18	19	20
21	22	23	24	25	26	27
28	29					

MARCH
S	M	T	W	T	F	S
		1	2	3	4	5
6	7	8	9	10	11	12
13	14	15	16	17	18	19
20	21	22	23	24	25	26
27	28	29	30	31		

APRIL
S	M	T	W	T	F	S
					1	2
3	4	5	6	7	8	9
10	11	12	13	14	15	16
17	18	19	20	21	22	23
24	25	26	27	28	29	30

MAY
S	M	T	W	T	F	S
1	2	3	4	5	6	7
8	9	10	11	12	13	14
15	16	17	18	19	20	21
22	23	24	25	26	27	28
29	30	31				

JUNE
S	M	T	W	T	F	S
			1	2	3	4
5	6	7	8	9	10	11
12	13	14	15	16	17	18
19	20	21	22	23	24	25
26	27	28	29	30		

M

JULY

S	M	T	W	T	F	S
					1	2
3	4	5	6	7	8	9
10	11	12	13	14	15	16
17	18	19	20	21	22	23
24	25	26	27	28	29	30
31						

AUGUST

S	M	T	W	T	F	S
	1	2	3	4	5	6
7	8	9	10	11	12	13
14	15	16	17	18	19	20
21	22	23	24	25	26	27
28	29	30	31			

SEPTEMBER

S	M	T	W	T	F	S
				1	2	3
4	5	6	7	8	9	10
11	12	13	14	15	16	17
18	19	20	21	22	23	24
25	26	27	28	29	30	

OCTOBER

S	M	T	W	T	F	S
						1
2	3	4	5	6	7	8
9	10	11	12	13	14	15
16	17	18	19	20	21	22
23	24	25	26	27	28	29
30	31					

NOVEMBER

S	M	T	W	T	F	S
		1	2	3	4	5
6	7	8	9	10	11	12
13	14	15	16	17	18	19
20	21	22	23	24	25	26
27	28	29	30			

DECEMBER

S	M	T	W	T	F	S
				1	2	3
4	5	6	7	8	9	10
11	12	13	14	15	16	17
18	19	20	21	22	23	24
25	26	27	28	29	30	31

N 1780 1820 1848 1876 1916
   1944 1972 2000 2028

JANUARY

S	M	T	W	T	F	S
						1
2	3	4	5	6	7	8
9	10	11	12	13	14	15
16	17	18	19	20	21	22
23	24	25	26	27	28	29
30	31					

FEBRUARY

S	M	T	W	T	F	S
		1	2	3	4	5
6	7	8	9	10	11	12
13	14	15	16	17	18	19
20	21	22	23	24	25	26
27	28	29				

MARCH

S	M	T	W	T	F	S
			1	2	3	4
5	6	7	8	9	10	11
12	13	14	15	16	17	18
19	20	21	22	23	24	25
26	27	28	29	30	31	

APRIL

S	M	T	W	T	F	S
						1
2	3	4	5	6	7	8
9	10	11	12	13	14	15
16	17	18	19	20	21	22
23	24	25	26	27	28	29
30						

MAY

S	M	T	W	T	F	S
	1	2	3	4	5	6
7	8	9	10	11	12	13
14	15	16	17	18	19	20
21	22	23	24	25	26	27
28	29	30	31			

JUNE

S	M	T	W	T	F	S
				1	2	3
4	5	6	7	8	9	10
11	12	13	14	15	16	17
18	19	20	21	22	23	24
25	26	27	28	29	30	

N

JULY
S	M	T	W	T	F	S
						1
2	3	4	5	6	7	8
9	10	11	12	13	14	15
16	17	18	19	20	21	22
23	24	25	26	27	28	29
30	31					

AUGUST
S	M	T	W	T	F	S
		1	2	3	4	5
6	7	8	9	10	11	12
13	14	15	16	17	18	19
20	21	22	23	24	25	26
27	28	29	30	31		

SEPTEMBER
S	M	T	W	T	F	S
					1	2
3	4	5	6	7	8	9
10	11	12	13	14	15	16
17	18	19	20	21	22	23
24	25	26	27	28	29	30

OCTOBER
S	M	T	W	T	F	S
1	2	3	4	5	6	7
8	9	10	11	12	13	14
15	16	17	18	19	20	21
22	23	24	25	26	27	28
29	30	31				

NOVEMBER
S	M	T	W	T	F	S
			1	2	3	4
5	6	7	8	9	10	11
12	13	14	15	16	17	18
19	20	21	22	23	24	25
26	27	28	29	30		

DECEMBER
S	M	T	W	T	F	S
					1	2
3	4	5	6	7	8	9
10	11	12	13	14	15	16
17	18	19	20	21	22	23
24	25	26	27	28	29	30
31						

Chinese calendar
The Chinese New Year starts on the day of the first
new moon after the sun enters Aquarius (between 21
January and 19 February)

Dog	Chicken	Monkey	Sheep	Horse	Snake
1910	1909	1908	1907	1906	1905
1922	1921	1920	1919	1918	1917
1934	1933	1932	1931	1930	1929
1946	1945	1944	1943	1942	1941
1958	1957	1956	1955	1954	1953
1970	1969	1968	1967	1966	1965
1982	1981	1980	1979	1978	1977
1994	1993	1992	1991	1990	1989
2006	2005	2004	2003	2002	2001

Dragon	Rabbit	Tiger	Ox	Rat	Pig
1904	1903	1902	1901	1900	1911
1916	1915	1914	1913	1912	1923
1928	1927	1926	1925	1924	1935
1940	1939	1938	1937	1936	1947
1952	1951	1950	1949	1948	1959
1964	1963	1962	1961	1960	1971
1976	1975	1974	1973	1972	1983
1988	1987	1986	1985	1984	1995
2000	1999	1998	1997	1996	2007

The zodiac year

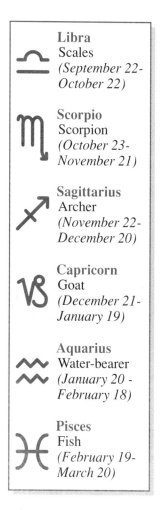

Aries
Ram
*(March 21-
April 20)*

Taurus
Bull
*(April 21-
May 20)*

Gemini
Twins
*(May 21-
June 20)*

Cancer
Crab
*(June 21-
July 21)*

Leo
Lion
*(July 22-
August 21)*

Virgo
Virgin
*(August 22-
September 21)*

Libra
Scales
*(September 22-
October 22)*

Scorpio
Scorpion
*(October 23-
November 21)*

Sagittarius
Archer
*(November 22-
December 20)*

Capricorn
Goat
*(December 21-
January 19)*

Aquarius
Water-bearer
*(January 20 -
February 18)*

Pisces
Fish
*(February 19-
March 20)*

Time zones of the world
Some countries, including the UK, adopt Daylight
Saving Time (DST) in order to receive more daylight in

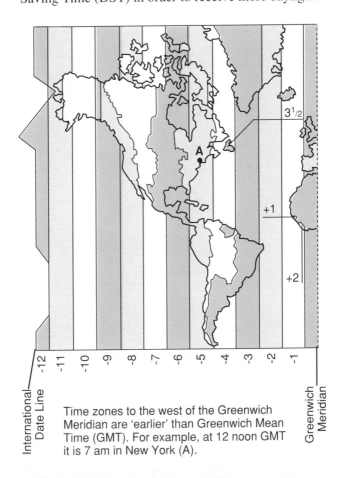

Time zones to the west of the Greenwich
Meridian are 'earlier' than Greenwich Mean
Time (GMT). For example, at 12 noon GMT
it is 7 am in New York (A).

summer. Clocks are put forward 1 hour in spring and back 1 hour in autumn. The maps below do not reflect DST adjustments.

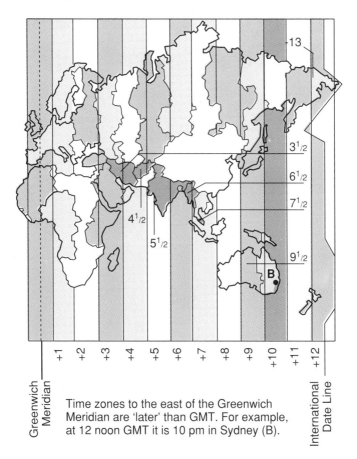

Time zones to the east of the Greenwich Meridian are 'later' than GMT. For example, at 12 noon GMT it is 10 pm in Sydney (B).

Comparative times

Instead of everyone telling the time by the Sun, a
system of time zones has been introduced. All the
places in the same zone share the same time.
Worldwide there are 24 zones, one for each hour of the
day and night. The following list gives the time in cities
around the world when it is 12 noon at Greenwich.

Adelaide	9.30 pm	Buenos Aires	9 am
Addis Ababa	3 pm	Cairo	2 pm
Alexandria	2 pm	Calcutta	5.30 pm
Algiers	1 pm	Calgary	5 am
Amsterdam	1 pm	Cape Town	2 pm
Ankara	2 pm	Caracas	8 am
Athens	2 pm	Casablanca	12 noon
Baghdad	3 pm	Chicago	6 am
Bangkok	12 midnight	Colombo	5.30 pm
Barcelona	1 pm	Copenhagen	1 pm
Beijing	8 pm	Dakar	11 am
Belfast	12 noon	Delhi	5.30 pm
Belgrade	1 pm	Dublin	12 noon
Berlin	1 pm	Edinburgh	12 noon
Bogotá	7 am	Florence	1 pm
Bombay	5.30 pm	Frankfurt	1 pm
Boston	7 am	Geneva	1 pm
Brasilia	9 pm	Gibraltar	1 pm
Brussels	1 pm	Glasgow	12 noon
Bucharest	2 pm	Guangzhou	8 pm
Budapest	1 pm	Guatemala City	6 am

Guayaquil	7 am	Mexico City	6 am
Halifax, Nova Scotia	8 am	Monaco-Ville	1 pm
Hanoi	8 pm	Montevideo	9 am
Havana	12 noon	Montreal	7 am
Helsinki	7 pm	Moscow	3 pm
Hobart	10 pm	Munich	1 pm
Ho Chi Minh City	8 pm	Nairobi	3 pm
Hong Kong	8 pm	Naples	1 pm
Honolulu	2 am	New Orleans	6 am
Istanbul	2 pm	New York	7 am
Jakarta	12 midnight	Oslo	1 pm
Jerusalem	2 pm	Ottawa	7 am
Johannesburg	2 pm	Panama	7 am
Karachi	5 pm	Paris	1 pm
Kingston, Jamaica	7 am	Perth, Australia	8 pm
Kuala Lumpur	8 pm	Prague	1 pm
La Paz	8 am	Quebec	7 am
Leningrad	3 pm	Rangoon	6.30 pm
Leopoldville	1 pm	Reykjavik	11 am
Lima	7 am	Rio de Janeiro	9 am
Lisbon	12 noon	Riyadh	3 pm
Liverpool	12 noon	Rome	1 pm
London	12 noon	San Francisco	4 am
Madrid	1 pm	San Juan	8 am
Managua	6 am	Santiago	8 am
Manila	8 pm	Seoul	9 pm
Marseilles	1 pm	Shanghai	8 pm
Mecca	3 pm	Singapore	7.30 pm
Melbourne	10 pm	Sofia	2 pm

Stockholm	1 pm	Venice	1 pm
Sydney	10 pm	Vienna	1 pm
Tangiers	12 noon	Vladivostock	10 pm
Teheran	3.30 pm	Warsaw	1 pm
Tel Aviv	2 pm	Wellington	12 midnight
Tokyo	9 pm	Winnipeg	6 am
Toronto	7 am	Yokohama	9 pm
Tripoli	2 pm	Zurich	1 pm
Vancouver	4 am		

Time zones: United States
Apart from Alaska and Hawaii, U.S. states and
Washington, DC are divided into four time zones:
Eastern, Central, Mountain, and Pacific. The time in
each zone is one hour earlier than in the zones to its
east and one hour later in the zone to the west. The
basic pattern of time zones in states is given below.
* States which fall into two time zones

---

1 Eastern (12 noon)

Connecticut	Maine	Pennsylvania
Delaware	Maryland	Rhode Island
District of	Massachusetts	South Carolina
Columbia	Michigan*	Tennessee*
Florida*	New Hampshire	Vermont
Georgia	New Jersey	Virginia
Indiana*	North Carolina	West Virginia
Kentucky*	Ohio	

2 Central (11 am)

Alabama	Kentucky*	North Dakota*
Arkansas	Louisiana	Oklahoma
Florida*	Michigan*	South Dakota
Illinois	Minnesota	Tennessee*
Indiana*	Mississippi	Texas*
Iowa	Missouri	Wisconsin
Kansas*	Nebraska*	

3 Mountain (10 am)

Arizona	Nebraska*	Texas*
Colorado	New Mexico	Utah
Idaho*	North Dakota*	Wyoming
Kansas*	Oregon*	
Montana	South Dakota*	

4 Pacific (9 am)

California	Nevada	Washington
Idaho*	Oregon*	

Office times The table shows the usual office hours
(local time) in various cities compared with UK time.

					AM					
UK time	3	4	5	6	7	8	9	10	11	12
London							09.00			
Sydney										
Brussels						08.30		12.00		
Rio de Janeiro										
Toronto										
Copenhagen					08.00					
Paris						09.00		12.00		
Athens				08.00					14.00	
Milan					08.30			12.45		
Tokyo						09.00				
Amsterdam					08.30					
Oslo				08.00						
Lisbon							10.00	12.30		
Dublin						09.30				
Riyadh				08.00			13.00			
Johannesburg					08.30					
Madrid					09.30		13.30			
Stockholm					08.30					
Geneva				08.00			12.00			
New York										
Frankfurt					08.00					

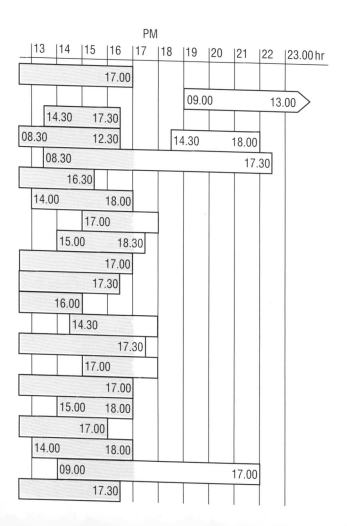

Wedding anniversaries

Year	Traditional (alternative)	Modern
1st	Paper	Clocks
2nd	Cotton	China
3rd	Leather	Crystal, glass
4th	Linen (silk)	Electrical appliances
5th	Wood	Silverware
6th	Iron	Wood
7th	Wool (copper)	Desk sets
8th	Bronze	Linen, lace
9th	Pottery (china)	Leather
10th	Tin (aluminium)	Diamond jewellery
11th	Steel	Fashion jewellery, accessories
12th	Silk	Pearls or coloured gems
13th	Lace	Textile, furs
14th	Ivory	Gold jewellery
15th	Crystal	Watches
20th	China	Platinum
25th	Silver	Sterling silver jubilee
30th	Pearl	Diamond
35th	Coral (jade)	Jade
40th	Ruby	Ruby
45th	Sapphire	Sapphire
50th	Gold	Gold
55th	Emerald	Emerald
60th	Diamond	Diamond

Geological timescale

Years ago (millions)	Epoch	Period	Era
.01	Holocene	Quaternary	Cenozoic
2	Pleistocene	Quaternary	Cenozoic
5	Pliocene	Tertiary	Cenozoic
25	Miocene	Tertiary	Cenozoic
38	Oligocene	Tertiary	Cenozoic
55	Eocene	Tertiary	Cenozoic
65	Palaeocene	Tertiary	Cenozoic
145		Cretaceous	Mesozoic
215		Jurassic	Mesozoic
250		Triassic	Mesozoic
285		Permian	Palaeozoic
360		Carboniferous	Palaeozoic
410		Devonian	Palaeozoic
440		Silurian	Palaeozoic
500		Ordovician	Palaeozoic
600		Cambrian	Palaeozoic
1000		Proterozoic	Precambrian
4600		Archaeozoic	Precambrian

8: Speed

Formulas
Below are listed the multiplication/division factors for converting units of speed from imperial to metric, and vice versa; and, also, for converting from one unit to another within the same system. Note that two kinds of factors are given: quick, for an approximate conversion that can be made without a calculator; and accurate, for an exact conversion.

			Quick	**Accurate**
Miles per hour (mph)				
Kilometres per hour (km/h)				
mph	⟶	km/h	× 1.5	× 1.609
km/h	⟶	mph	÷ 1.5	× 0.621
Yards per minute (ypm)				
Metres per minute (m/min)				
ypm	⟶	m/min	× 1	× 1.094
m/min	⟶	ypm	÷ 1	× 0.914
Metres per minute (m/min)				
Feet per minute (ft/min)				
m/min	⟶	ft/min	× 3	× 3.281
ft/min	⟶	m/min	÷ 3	× 0.305
Inches per second (in/s)				
Centimetres per second (cm/s)				
in/s	⟶	cm/s	× 2.5	× 2.54
cm/s	⟶	in/s	÷ 2.5	× 0.394

	International knots (kn) Miles per hour (mph)	**Quick**	**Accurate**
	kn ⟶ mph	× 1	× 1.151
	mph ⟶ kn	÷ 1	× 0.869

	British knots (UK kn) International knots (kn)		
	UK kn ⟶ kn	× 1	× 1.001
	kn ⟶ UK kn	÷ 1	× 0.999

	International knots (kn) Kilometres per hour (km/h)		
	kn ⟶ km/h	× 2	× 1.852
	km/h ⟶ kn	÷ 2	× 0.540

	Miles per hour (mph) Feet per second (ft/s)		
	mph ⟶ ft/s	× 1.5	× 1.467
	ft/s ⟶ mph	÷ 1.5	× 0.682

	Metres per second (m/s) Kilometres per hour (km/h)		
	m/s ⟶ km/h	× 3.5	× 3.599
	km/h ⟶ m/s	÷ 3.5	× 0.278

Conversion tables

The tables below can be used to convert units of speed
from one measuring system to another. The first group

Miles per hour to Kilometres per hour		Kilometres per hour to Miles per hour		Yards per minute to Metres per minute	
mph	km/h	km/h	mph	ypm	m/min
1	1.609	1	0.621	1	0.914
2	3.219	2	1.242	2	1.829
3	4.828	3	1.864	3	2.743
4	6.437	4	2.485	4	3.658
5	8.047	5	3.106	5	4.572
6	9.656	6	3.728	6	5.486
7	11.265	7	4.349	7	6.401
8	12.875	8	4.970	8	7.315
9	14.484	9	5.592	9	8.230
10	16.093	10	6.213	10	9.144
20	32.187	20	12.427	20	18.288
30	48.280	30	18.641	30	27.432
40	64.374	40	24.854	40	36.576
50	80.467	50	31.068	50	45.720
60	96.561	60	37.282	60	54.864
70	112.654	70	43.495	70	64.008
80	128.748	80	49.709	80	73.152
90	144.841	90	55.923	90	82.296
100	160.934	100	62.137	100	91.440

of tables converts imperial to metric, and vice versa.
The tables beginning on page 192 convert knots,
imperial and metric units.

Metres per minute to Yards per minute		Feet per minute to Metres per minute		Metres per minute to Feet per minute	
m/min	ypm	ft/min	m/min	m/min	ft/min
1	1.094	1	0.305	1	3.281
2	2.187	2	0.610	2	6.562
3	3.281	3	0.914	3	9.842
4	4.374	4	1.219	4	13.123
5	5.468	5	1.524	5	16.404
6	6.562	6	1.829	6	19.685
7	7.655	7	2.134	7	22.966
8	8.749	8	2.438	8	26.246
9	9.842	9	2.743	9	29.527
10	10.936	10	3.048	10	32.808
20	21.872	20	6.096	20	65.616
30	32.808	30	9.144	30	98.424
40	43.744	40	12.192	40	1.1.232
50	54.680	50	15.240	50	164.040
60	65.616	60	18.288	60	196.848
70	76.552	70	21.336	70	229.656
80	87.488	80	24.384	80	262.464
90	98.424	90	27.432	90	295.272
100	109.360	100	30.480	100	328.080

Imperial and metric units of speed (continued)

Inches per second to Centimetres per second		Centimetres per second to Inches per second		International knots to Miles per hour	
in/s	cm/s	cm/s	in/s	kn	mph
1	2.54	1	0.394	1	1.151
2	5.08	2	0.787	2	2.302
3	7.62	3	1.181	3	3.452
4	10.16	4	1.579	4	4.603
5	12.70	5	1.969	5	5.753
6	15.24	6	2.362	6	6.905
7	17.78	7	2.760	7	8.055
8	20.32	8	3.150	8	9.206
9	22.86	9	3.543	9	10.357
10	25.40	10	3.937	10	11.508
20	50.80	20	7.874	20	23.016
30	76.20	30	11.811	30	34.523
40	101.60	40	15.748	40	46.031
50	127.00	50	19.685	50	57.540
60	152.40	60	23.622	60	69.047
70	177.80	70	27.559	70	80.555
80	203.20	80	31.496	80	92.062
90	228.60	90	35.433	90	103.570
100	254.00	100	39.370	100	115.078

Miles per hour to International knots		UK knots to International knots		International knots to UK knots	
mph	kn	UK kn	kn	kn	UK kn
1	0.869	1	1.001	1	0.999
2	1.738	2	2.001	2	1.999
3	2.607	3	3.002	3	2.998
4	3.476	4	4.003	4	3.997
5	4.345	5	5.003	5	4.997
6	5.214	6	6.004	6	5.996
7	6.083	7	7.004	7	6.996
8	6.952	8	8.005	8	7.995
9	7.821	9	9.006	9	8.994
10	8.690	10	10.006	10	9.994
20	17.380	20	20.013	20	19.987
30	26.069	30	30.019	30	29.981
40	34.759	40	40.026	40	39.974
50	43.449	50	50.032	50	49.968
60	52.139	60	60.038	60	59.962
70	60.828	70	70.045	70	69.955
80	69.518	80	80.051	80	79.949
90	78.208	90	90.058	90	89.942
100	86.898	100	100.064	100	99.936

Imperial and metric units of speed (continued)

International knots to Kilometres per hour		Kilometres per hour to International knots		Miles per hour to Feet per second	
kn	km/h	km/h	kn	mph	ft/s
1	1.852	1	0.540	1	1.467
2	3.704	2	1.08	2	2.933
3	5.556	3	1.62	3	4.400
4	7.408	4	2.16	4	5.867
5	9.260	5	2.70	5	7.334
6	11.112	6	3.23	6	8.800
7	12.964	7	3.77	7	10.267
8	14.816	8	4.31	8	11.734
9	16.668	9	4.85	9	13.203
10	18.520	10	5.30	10	14.667
20	37.040	20	10.78	20	29.334
30	55.560	30	16.17	30	44.001
40	74.080	40	21.56	40	58.668
50	92.600	50	26.95	50	73.335
60	111.120	60	32.34	60	88.002
70	129.640	70	37.73	70	102.669
80	148.160	80	43.12	80	117.336
90	166.680	090	48.51	90	132.003
100	185.200	100	53.90	100	146.670

Feet per second to Miles per hour		Kilometres per hour to Metres per second		Metres per second to Kilometres per hour	
ft/s	mph	km/h	m/s	m/s	km/h
1	0.682	1	0.278	1	3.599
2	1.364	2	0.556	2	7.198
3	2.046	3	0.834	3	10.797
4	2.728	4	1.111	4	14.396
5	3.410	5	1.389	5	17.995
6	4.092	6	1.669	6	21.594
7	4.774	7	1.945	7	25.193
8	5.456	8	2.222	8	28.792
9	6.138	9	2.500	9	32.391
10	6.820	10	2.778	10	35.990
20	13.640	20	5.556	20	71.980
30	20.460	30	8.334	30	107.970
40	27.280	40	11.112	40	143.960
50	34.100	50	13.890	50	179.950
60	40.920	60	16.668	60	215.940
70	47.740	70	19.446	70	251.930
80	54.560	80	22.224	80	287.920
90	61.380	90	25.002	90	323.910
100	68.200	100	27.780	100	359.900

Wind speeds

Wind is the movement of air across the surface of the
Earth. Its direction is determined by a combination of
factors: the Earth's rotation, land features such as
mountains, and variations in the temperature and
pressure of the atmosphere.

Beaufort scale

Wind speed is measured using the internationally

Beaufort scale

km/h	10	20	30	40	50	60

Number	Description	Speed range	
		km/h	mph
0	Calm	Below 1	Below 1
1	Light air	1–5	1–3
2	Light breeze	6–12	4–7
3	Gentle breeze	13–20	8–12
4	Mederate breeze	21–29	13–18
5	Fresh breeze	30–39	19–24
6	Strong breeze	40–50	25–31
7	Moderate gale	51–61	32–38
8	Fresh gale	62–74	39–46
9	Strong gale	75–87	47–54
10	Whole gale	88–102	55–63
11	Storm	103–120	64–75
12–17	Hurricane	Over 120	Over 75

recognised Beaufort scale, named after the nineteenth-century British admiral, Sir Francis Beaufort.

Below are listed the Beaufort wind force numbers and the range of speeds to which they apply. The brief official descriptions show the variety of wind speeds that are measured, and examples of their physical manifestations are given to illustrate the strengths of their relative forces.

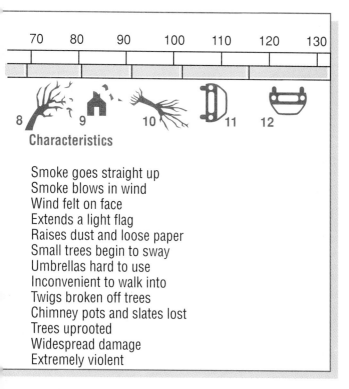

70   80   90   100   110   120   130

8   9   10   11   12

Characteristics

Smoke goes straight up
Smoke blows in wind
Wind felt on face
Extends a light flag
Raises dust and loose paper
Small trees begin to sway
Umbrellas hard to use
Inconvenient to walk into
Twigs broken off trees
Chimney pots and slates lost
Trees uprooted
Widespread damage
Extremely violent

9: Geometry

Polygons

Name of polygon	Number of sides	Each internal angle	Sum of internal angles
Triangle	3	60°	180°
Square	4	90°	360°
Pentagon	5	108°	540°
Hexagon	6	120°	720°
Heptagon	7	128.6°	900°
Octagon	8	135°	1080°
Nonagon	9	140°	1260°
Decagon	10	144°	1440°
Undecagon	11	147.3°	1620°
Dodecagon	12	150°	1800°

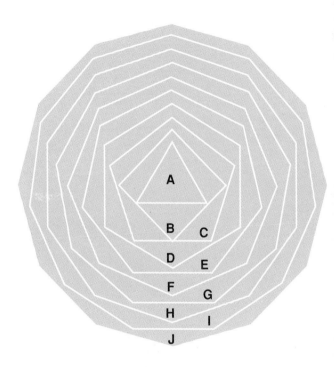

A Triangle	**F** Octagon
B Square	**G** Nonagon
C Pentagon	**H** Decagon
D Hexagon	**I** Undecagon
E Heptagon	**J** Dodecagon

Quadrilaterals

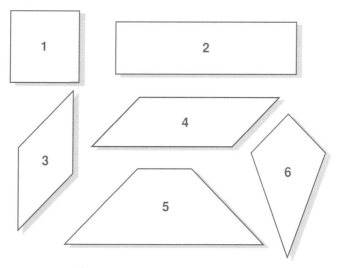

Quadrilaterals

A quadrilateral is a four-sided polygon.

1 Square		All the sides are the same length and all the angles are right angles.
2 Rectangle		Opposite sides are the same length and all the angles are right angles.
3 Rhombus		All the sides are the same length but none of the angles are right angles.
4 Parallelogram		Opposite sides are parallel to each other and of the same length.
5 Trapezium		One pair of the opposite sides is parallel.
6 Kite		Adjacent sides are the same length and the diagonals intersect at right angles.

Triangles

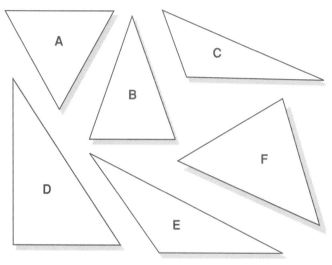

Triangles

A Equilateral	All the sides are the same length and all the angles are equal.	
B Isosceles	Two sides are of the same length and two angles are of equal size.	
C Scalene	All the sides are of different length and all the angles are of different sizes.	
D Right angle	A triangle that contains one right angle.	
E Obtuse angle	A triangle that contains one obtuse angle.	
F Acute angle	A triangle with three acute angles.	

10: Everyday measures

Standard UK paper sizes

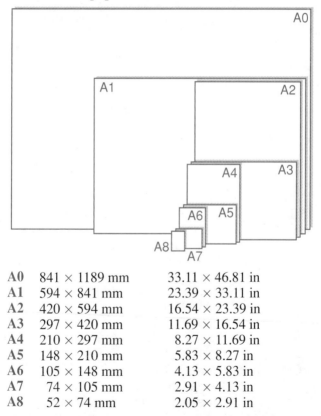

A0	841 × 1189 mm	33.11 × 46.81 in
A1	594 × 841 mm	23.39 × 33.11 in
A2	420 × 594 mm	16.54 × 23.39 in
A3	297 × 420 mm	11.69 × 16.54 in
A4	210 × 297 mm	8.27 × 11.69 in
A5	148 × 210 mm	5.83 × 8.27 in
A6	105 × 148 mm	4.13 × 5.83 in
A7	74 × 105 mm	2.91 × 4.13 in
A8	52 × 74 mm	2.05 × 2.91 in

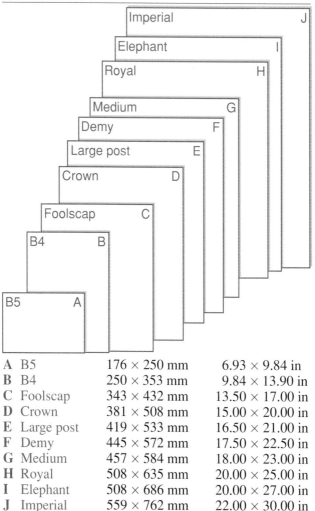

A	B5	176 × 250 mm	6.93 × 9.84 in
B	B4	250 × 353 mm	9.84 × 13.90 in
C	Foolscap	343 × 432 mm	13.50 × 17.00 in
D	Crown	381 × 508 mm	15.00 × 20.00 in
E	Large post	419 × 533 mm	16.50 × 21.00 in
F	Demy	445 × 572 mm	17.50 × 22.50 in
G	Medium	457 × 584 mm	18.00 × 23.00 in
H	Royal	508 × 635 mm	20.00 × 25.00 in
I	Elephant	508 × 686 mm	20.00 × 27.00 in
J	Imperial	559 × 762 mm	22.00 × 30.00 in

Envelope sizes and styles

Three of the most popular envelope styles are
illustrated on the opposite page. There are many
variations of each of these styles.

Below are illustrated the primary envelope sizes; the
table opposite gives dimensions and the styles in which
each is available.

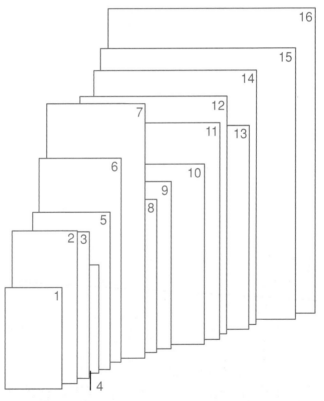

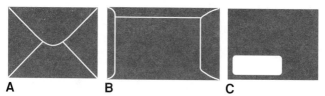

A Banker style
B Pocket style
C Window

Number	inches	mm	A	B	C
1	$3\frac{1}{2}$ x 6	89 x 152	•	•	•
2	4 x 9	102 x 229	•	•	•
3	$4\frac{1}{4}$ x $8\frac{5}{8}$	110 x 220	•	•	•
4	$4\frac{1}{2}$ x $6\frac{3}{8}$	114 x 162	•		•
5	$4\frac{3}{4}$ x $9\frac{1}{4}$	120 x 235	•	•	
6	5 x 12	127 x 305		•	
7	6 x 15	152 x 381		•	
8	$6\frac{3}{8}$ x 9	162 x 229	•	•	•
9	$6\frac{7}{8}$ x $9\frac{7}{8}$	175 x 250		•	
10	$8\frac{1}{2}$ x $10\frac{5}{8}$	216 x 270		•	
11	9 x $12\frac{3}{4}$	229 x 324		•	•
12	9 x 14	229 x 356		•	
13	10 x 12	254 x 305		•	
14	10 x 15	254 x 381		•	
15	12 x 16	305 x 406		•	
16	$12\frac{3}{4}$ x 18	324 x 457		•	

Book sizes

A	Collins Pocket Reference	157 × 112 mm	6.20 × 4.40 in
B	Foolscap octavo	170 × 110 mm	6.75 × 4.25 in
C	Crown octavo	190 × 125 mm	7.50 × 5.00 in
D	Large crown octavo	205 × 135 mm	8.00 × 5.25 in
E	Small demy octavo	215 × 145 mm	8.50 × 5.63 in
F	Demy octavo	220 × 145 mm	8.75 × 5.63 in
G	Medium octavo	230 × 145 mm	9.00 × 5.75 in
H	Small royal octavo	235 × 155 mm	9.25 × 6.13 in
I	Royal octavo	255 × 160 mm	10.00 × 6.25 in

Beverage measures

Beer measures
1 nip = $^1/_4$ pint
1 small = $^1/_2$ pint
1 large = 1 pint
1 flagon = 1 quart
1 anker = 10 gallons
1 firkin = 9 gallons

1 kilderkin = 2 firkins
1 barrel = 2 kilderkins
1 hogshead = $1^1/_2$ barrels
1 butt = 2 hogsheads
1 tun = 2 butts
        216 gallons

Handy measures
small jigger = 1 fl oz
small wine
  glass = 2 fl oz
cocktail glass = $^1/_4$ pint
sherry glass = $^1/_4$ pint
large wine
  glass = $^1/_4$ pint
tumbler = $^1/_2$ pint

Wine measures
10 gallons = 1 anker
1 hogshead = 63 gallons
2 hogsheads = 1 pipe
2 pipes = 1 tun
1 puncheon = 84 gallons
1 butt
 (sherry) = 110 gallons

US spirits measures
1 pony = $^1/_2$ jigger
1 jigger = $1^1/_2$ shot
1 shot = 1 fl oz
1 pint = 16 shots
1 fifth = 25.6 shots
        1.6 pints
        0.8 quart
        0.758 litre

1 quart = 32 shots
        $1^1/_4$ fifths
1 magnum
of wine = 2 quarts
        $2^1/_2$ bottles

Standard bottle content

The number of glasses in a bottle is usually based on pub
measures: wine glasses hold an average 113 ml/4 fl oz,
sherry glasses 50 ml/$\frac{1}{3}$ gill. Spirit measures are 25 ml/$\frac{1}{6}$ gill
in England and Wales, 30 ml/$\frac{1}{5}$ gill in Scotland. The chart
(below) shows how many measures there are in a bottle.

Bottle	Content of bottle	Glasses and measures
Wine	75 cl	6$\frac{1}{2}$ glasses of 113 ml
Sherry	75 cl	15 glasses of 50 ml
Spirits	70 cl	28 glasses of 25 ml

Average alcoholic content

To compare the alcoholic strength of different drinks, a system of standard units has been introduced. The table (below) is intended as a rough guide only, and the exact number of units per measure varies with the strength of the drink.

Beverage	Quantity	Units of alcohol
Beer, lager, cider	1 pint	2
Spirits: vodka, rum, whisky, gin, etc.	Single measure 25 ml/$\frac{1}{6}$ gill	1
Wine	Standard glass 113 ml/4 fl oz	1
Fortified wines: sherry, madeira port, etc.	Small glass 50 ml/$\frac{1}{3}$ gill	1

Alcoholic units

The chart (below) compares alcoholic content for
different kinds of drink. Each category shows the
alcoholic content by volume and the number of standard

Beverage	Quantity	% Alcohol	Units
Spirits	70 cl bottle		
		43	30
		40	28
		38	27
		37	26
Fortified wines	75 cl bottle		
		27	20
		25	19
		23	$17\frac{1}{2}$
		20	15
		17	13
		15	$11\frac{1}{2}$
Wine	75 cl bottle		
		15	11
		13	10
		11	8
		9	7
		7	$5\frac{1}{2}$
Beers and wine	75 cl bottle		
		6	$4\frac{1}{2}$
		5	4
		3	2
Low alcohol			
		1	1
		0.5	$1\frac{1}{2}$
		0.05	$1\frac{1}{25}$

units in typical bottles of given sizes. Normally, alcoholic content is expressed as a percentage on the label of the bottle. The maximum recommended number of units per week is 21 for men and 14 for women.

Beverage	Quantity	Calories
Beers, lagers and cider	½ pint (284 ml/10 fl oz)	
Sweet cider		110
Dry cider		95
Bitter		90
Ordinary strength lager		85
Brown ale		80
Light or mild ale		70
Low-alcohol lager		60
Spirits	1 pub measure (25 ml/⅙ gin)	
Brandy, whisky, gin, rum or vodka		50
Wine	An average glass (113 ml/4 fl oz)	
Sweet white		100
Rosé		85
Dry, white or red		75
Sherry	1 pub measure (50 ml/⅓ gill)	
Cream		70
Medium		60
Dry		55
Soft drinks		
Can of coke		130
Glass of orange juice		80
Ordinary tonic water		35
Low calorie tonic water		0
Diet coke		0

Champagne bottle sizes

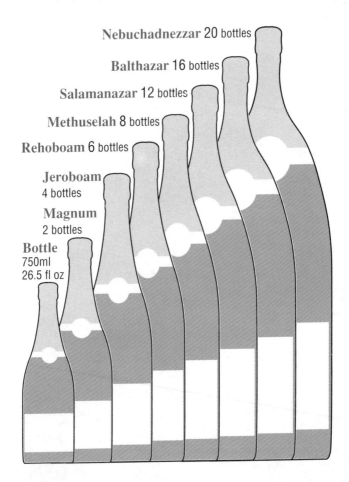

Nebuchadnezzar 20 bottles

Balthazar 16 bottles

Salamanazar 12 bottles

Methuselah 8 bottles

Rehoboam 6 bottles

Jeroboam
4 bottles

Magnum
2 bottles

Bottle
750ml
26.5 fl oz

Wine bottle shapes

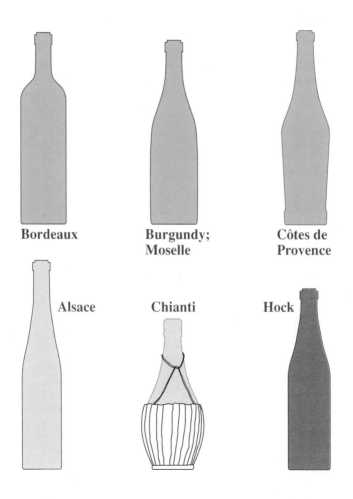

Bordeaux

**Burgundy;
Moselle**

**Côtes de
Provence**

Alsace **Chianti** **Hock**

Cooking measures

Although the names of the units are often the same, US measures are slightly different from UK imperial measures – for example, the US pint is 16 ounces, and the UK imperial pint is 20 ounces. US cooks use different measures for liquids and solids; in the imperial system used in the UK, a fluid ounce is equal to a dry ounce. On average, US units are roughly 4/5 the size of UK units. Metric measures are rarely used for cooking in the US, except millilitres for small liquid amounts.

UK liquid and dry measures

60 minims = 1 dram
8 drams = 1 fl oz
5 fl oz = 1 gill
1 gill = 1/4 pint
1 pint = 20 fl oz
2 pints = 1 quart

4 quarts = 1 gallon
1 gallon = 10 lb (weight in water)
2 gallons = 1 peck
4 pecks = 1 bushel
36 bushels = 1 chaldron

US liquid measures

60 minims = 1 fl dram
8 fl drams = 1 fl oz
4 fl oz = 1 gill
4 gills = 1 pint
2 pints = 1 quart
4 quarts = 1 gallon

US dry measures

1 dry pint = 1/2 dry quart
2 dry pints = 1 dry quart
8 dry quarts = 1 peck
4 pecks = 1 bushel

Water weights

1 fl oz water = 1 oz
1 pint water = 1 1/4 lb

1 quart water = 2 1/2 lb
1 gallon water = 10 lb

Handy measures

Object	Imperial	Metric
1 thimbleful	30 drops	2.5 ml
60 drops	1 teaspoon	5 ml
1 teaspoon	1 dram	5 ml
1 dessertspoon	2 drams	10 ml
1 tablespoon	4 drams	20 ml
2 tablespoons	1 fl oz	40 ml
1 wine glass	2 fl oz	100 ml
1 tea cup	5 fl oz (1 gill)	200 ml
1 mug	10 fl oz (1/2 pint)	400 ml

American measures

1 US cup	8 fl oz	227 ml
1 US teaspoon	$\frac{1}{6}$ fl oz	5 ml
1 US tablespoon	$\frac{1}{2}$ fl oz	14 ml
1 US pint	16 fl oz	
1 US bushel	64 US pt	32.5 litres
1 US peck	16 US pt	8.8 litres
1 US fl oz		0.296 litres
1 US CWT	89 UK cwt	45.4 kg
1 US ton	89 UK ton	907 kg

Oven temperatures

Below is a table of Fahrenheit/Celsius conversions for common oven temperatures.

°F	°C	**Oven**
225	110	very cool
250	130	
275	140	cool
300	150	
325	170	moderate
350	180	
375	190	moderately hot
400	200	
425	220	hot
450	230	
475	240	very hot

For other conversions, use the following formulas:

°F to °C	Subtract 32, then divide by 945 (which equals 1.8).
°C to °F	Multiply by 945 (which equals 1.8), then add 32.

Laundry codes

Most garments contain a label giving laundering instructions, usually shown in terms of symbols, that tell you if any item is washable (or should be dry-cleaned) and how to wash it. The codes are listed below.

The table on the following pages lists the old and new codes, recommended temperatures (for machine- or hand-washing), and other machine settings, and the types of fabric that should be washed according to that code.

A Machine or hand wash
B Can be bleached
C Do not bleach
D Iron
E Do not iron

F Dry cleanable
G Do not dry clean
H Tumble dry
I Do not tumble dry

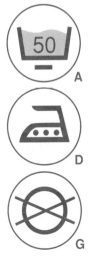

OLD	NEW	MACHINE WASH	HAND WASH
CODE		TEMPERATURE	
1 **9** **95** **95**	95	Very hot 95 °C to boil	Hand hot 50 °C or boil
2 **3** **60** **60**	60	Hot 60 °C	Hand hot 50 °C
4 **50**	50	Hand hot 50 °C	Hand hot 50 °C
5 **40**	40	Warm 40 °C	Warm 40 °C
6 **40**	40	Warm 40 °C	Warm 40 °C
7 **40**	40	Warm 40 °C	Warm 40 °C
8 **30**	30 30	Cool 30 °C	Cool 30 °C

AGITATION	RINSE	SPIN	FABRIC
Maximum	Normal	Normal	White cotton and linen with no special finish
Maximum	Normal	Normal	Cotton, linen, viscose, colour-fast with no special finish
Medium	Cold	Short spin or drip dry	Coloured nylon, polyester, cotton and viscose with special finish
Maximum	Normal	Normal	Cotton, linen, viscose, colour-fast to 40 °C
Minimum	Cold	Short spin	Acrylics, acetate and mixtures with wool
Minimum; do not rub	Normal	Normal; do not hand wring	Wool and wool mixtures
Minimum	Cold	Short spin; do not hand wring	Silk and printed acetate, not colour-fast at 40 °C

Clothing sizes

UK clothing sizes are equal to US sizes for some items, such as children's shoes; for others, the two vary slightly. Below are listed the European equivalents of UK and US clothing and shoe sizes. Remember also that sizes vary depending on the manufacturer.

Men's shoes

UK	USA	Europe
$6^1/_2$	7	39
7	$7^1/_2$	40
$7^1/_2$	8	41
8	$8^1/_2$	42
$8^1/_2$	9	43
9	$9^1/_2$	43
$9^1/_2$	10	44
10	$10^1/_2$	44
$10^1/_2$	11	45

Women's shoes

UK	USA	Europe
$3^1/_2$	5	36
$4^1/_2$	6	37
$5^1/_2$	7	38
$6^1/_2$	8	39
$7^1/_2$	9	40

Children's shoes

UK/USA	Europe
0	15
1	17
2	18
3	19
4	20
$4^1/_2$	21
5	22
6	23
7	24
8	25
$8^1/_2$	26
9	27
10	28
11	29
12	30
$12^1/_2$	31
13	32

Men's suits/overcoats

UK/USA	Europe
36	46
38	48
40	50
42	52
44	54
46	56

Men's shirts

UK/USA	Europe
12	30–31
12$\frac{1}{2}$	32
13	33
13$\frac{1}{2}$	34–35
14	36
14$\frac{1}{2}$	37
15	38
15$\frac{1}{2}$	39–40
16	41
16$\frac{1}{2}$	42
17	43
17$\frac{1}{2}$	44–45

Men's socks

UK/USA	Europe
9	38–39
10	39–40
10$\frac{1}{2}$	40–41
11	41–42
11$\frac{1}{2}$	42–43

Women's clothing

UK	USA	Europe
8	6	36
10	8	38
12	10	40
14	12	42
16	14	44
18	16	46
20	18	48
22	20	50
24	22	52

Children's clothing

UK	USA	Europe
16–18	2	40–45
20–22	4	50–55
24–26	6	60–65
28–30	7	70–75
32–34	8	80–85
36–38	9	90–95

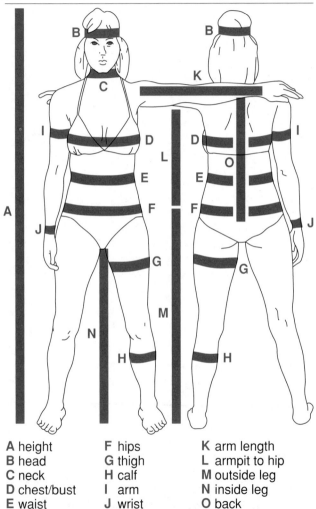

A height
B head
C neck
D chest/bust
E waist
F hips
G thigh
H calf
I arm
J wrist
K arm length
L armpit to hip
M outside leg
N inside leg
O back

Body measurements

The standard body measurements shown on the diagram on the opposite page are those needed for garment fitting.

Below are a few tips on taking some of these measurements.

Neck

Measure at the fullest part.

Chest/bust

Measure at the fullest part of the bust or chest and straight across the back.

Waist

Tie a string around the thinnest part of your body (the waist) and leave it there as a point of reference for other measurements.

Hips

There are two places to measure hips, depending on the garment: one is 2–4 in below the waist, at the top of the hipbones; the other is at the fullest part, usually 7–9 in below.

Arm

Measure at the fullest part, usually about 1 in below the armpit.

Arm length

Start at the shoulder bone and continue past the elbow to the wrist, with the arm slightly bent.

Back

Measure from the prominent bone in the back of the neck down the centre to the waist string.

Life expectancy
The table below shows life expectancy figures for
selected countries. Age in years appears at the top of
each bar.

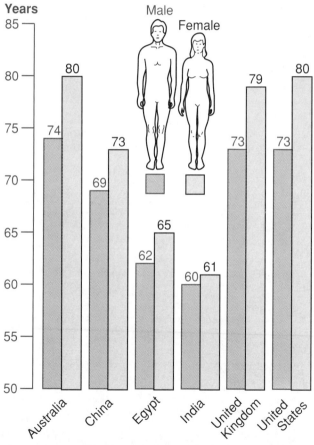

Average heights

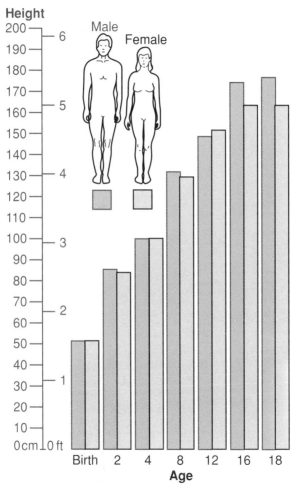

Average weights

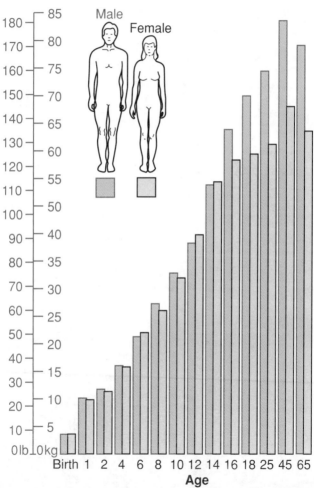

Wallpaper and tiling quantities

To calculate the amount of wallpaper needed to decorate a room, work out its surface area and divide the result by the surface area of one wallpaper roll. To work out the surface area of a room, multiply its circumference by its height (ignoring doors and windows unless they are very large, but taking account of skirting boards). Although wallpaper rolls are not absolutely standard, most are 53cm (21in) wide and 10m (33ft) long. Multiplying the length by the width gives the surface area.

Using the chart (below) you can estimate how many rolls of paper are needed to decorate a wide range of differently sized rooms.

Wall height (m)	Room circumference (m)												
	10	11	12	13	14	15	16	17	18	19	20	21	22
2.0–2.2	5	5	5	6	6	7	7	7	8	8	9	9	10
2.2–2.4	5	6	6	7	7	8	8	9	9	10	10	10	11
2.4–2.6	5	6	6	7	7	8	8	9	9	10	10	10	11
2.6–2.8	6	6	7	7	8	8	9	9	10	11	11	11	12
2.8–3.0	6	7	7	8	8	9	9	10	11	11	12	12	13
3.0–3.2	6	7	8	8	9	10	10	11	11	12	13	13	14
3.2–3.4	7	7	8	9	9	10	11	11	12	13	13	13	15

Classifying diamonds

As even a 1-carat diamond is scarcely part of everyday trade, diamonds are usually referred to as pointers, so you may be offered a '10 pointer' or '25 pointer'. A 10 pointer is one weighing 0.10 of a carat, and a 25 pointer weighs 0.25 of a carat.

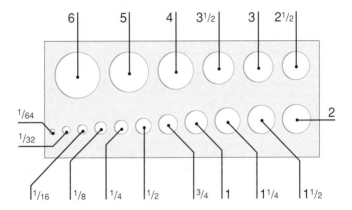

Occasionally jewellers sell diamonds by carat width, as shown in the gauge (above). So a stone which fits in the hole marked 3 could be referred to as '3 carats'. This is easily confused with weight, but a stone 3 carats wide may well be shallow and weigh much less than 3 carats. As the value of diamonds is by quality and weight – not width – make sure you ask what type of carat a jeweller is talking about and check the carat weight not just its width.

Classifying gemstones
Gems are classified by the form the crystals take –
aquamarines, for example, are hexagonal – and by the
way they refract light. Gems are measured in carats, a
measurement which goes back to keration, derived
from the Greek word for the carob beans which for
centuries were used as weights. But gem carats are
quite different from those used for gold. In gems a
carat is 200 mg ($^1/_5$ g or 3.086 grains troy weight) so
5 carats are 1 gram.
Knitting needle sizes
Here the British (metric) system is shown in colour,
and the American numbering system in black.

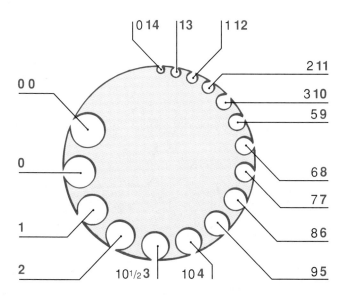

Gun gauge

A shotgun bore (diameter) is expressed in terms of gauge. Gauge was originally determined by the number of round lead balls – each the size of the shotgun bore – in a pound. For example, a 10-gauge shotgun was one that used balls that were 10 to the pound. The exception is the 410 bore, which is measured in inches: .410 in diameter, using 67.5 gauge. The most popular size today is the 12-gauge.

The table below shows gauge and equivalent bore size.

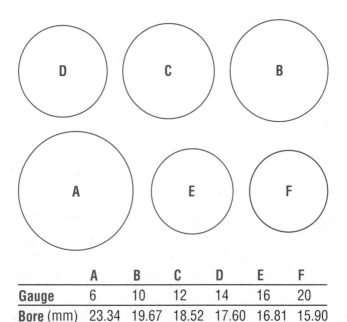

	A	B	C	D	E	F
Gauge	6	10	12	14	16	20
Bore (mm)	23.34	19.67	18.52	17.60	16.81	15.90

Horse measurements

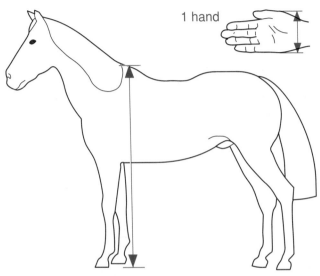

1 hand

The height of a horse or pony is measured to its withers (on the highest point on its back at the neck base), as shown above. Height is expressed in "hands high" (hh). One hand is 4 in (10 cm), the average width of a person's hand. Height is given to the nearest inch – a pony measuring 50 in (127 cm) is said to measure 12.2 hands. The table below shows recommended heights of ponies for young riders.

Pony's height (hh)	Child's age (years)
11–12	7–9
12–13	10–13
13–14.2	13–15
14.2–15.2	15–17

Odds in dice and cards
Dice

Odds in dice-throwing are determined by comparing favourable results with unfavourable. With one die, you have six possible results – one for each side of the die; with two die, you have 36 possible results. Some results – a 12 or a 2 - you have only one chance to achieve. Thus the odds against throwing a 12 or 2 are 35 to 1. For results with two possible combinations, the chances are 35 to 2, or 17 to 1. The table below shows the odds for each possible combination.

Combination	Chances	Combination
2	35–1	12
3	17–1	11
4	11–1	10
5	8.5–1	9
6	7–1	8
7	5–1	

Poker

Odds in poker are figured against a total number of possible combinations of 2 598 960. Thus, the odds of getting a royal flush (4 possible combinations) are 2 598 960 to 4, or 649 739 to 1.

Hand	Chances
royal flush	649 739 to 1
straight flush	72 192 to 1
four of a kind	4164 to 1
full house	693 to 1
flush	508 to 1
straight	254 to 1
three of a kind	46 to 1
two pairs	20 to 1
one pair	2.4 to 1
nil	2 to 1

Pontoon (Blackjack)

There are a possible 1326 combinations in pontoon; the odds of reaching 21 with two cards from a 52-card deck (64 possible combinations) are thus 1326 to 64, or 21 to 1.

Two-card total	Chances
21	21 to 1
20	9 to 1
19	16.5 to 1
18	15 to 1
17	14 to 1
16	15 to 1
15	14 to 1
14	13 to 1
13	11 to 1

11: Astronomy

Planetary features

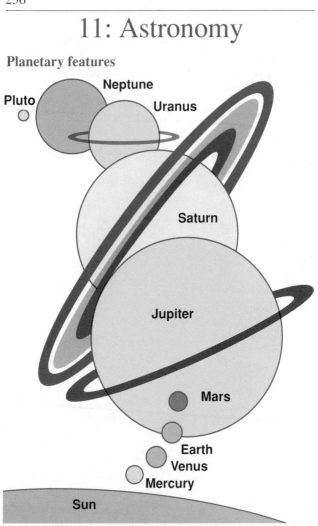

Diameter at equator

Planet	km	mi
Mercury	4878	2926.8
Venus	12 104	7262.4
Earth	12 756	7653.6
Mars	6795	4077.0
Jupiter	142 800	85 680.0
Saturn	120 000	72 000.0
Uranus	50 800	30 480.0
Neptune	48 500	29 100.0
Pluto	3000	1800.0

Rotation period

Mercury	58 days 15 hr
Venus	243 days
Earth	23 hr 56 min
Mars	24 hr 37 min
Jupiter	9 hr 50 min
Saturn	10 hr 14 min
Uranus	16 hr 10 min
Neptune	18 hr 26 min
Pluto	6 days 9 hr

Average surface temperatures

Solid surface		Cloud surface	
Mercury	350 °C (day) / −170 °C (night)	Jupiter	−150 °C
Venus	480 °C	Saturn	−180 °C
Earth	22 °C	Uranus	−210 °C
Mars	−23 °C	Neptune	−220 °C
		Pluto	−230 °C

Planetary distances

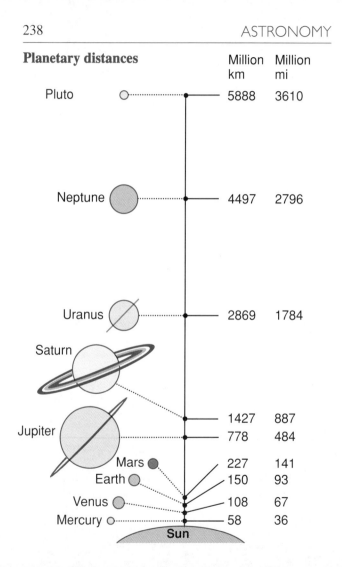

	Million km	Million mi
Pluto	5888	3610
Neptune	4497	2796
Uranus	2869	1784
Saturn	1427	887
Jupiter	778	484
Mars	227	141
Earth	150	93
Venus	108	67
Mercury	58	36

Sun

Mean distance from the Sun

Planet	km	mi
Mercury	58 000 000	36 000 000
Venus	108 000 000	67 000 000
Earth	150 000 000	93 000 000
Mars	227 000 000	141 000 000
Jupiter	778 000 000	484 000 000
Saturn	1 427 000 000	887 000 000
Uranus	2 869 000 000	1 784 000 000
Neptune	4 497 000 000	2 796 000 000
Pluto	5 888 000 000	3 661 000 000

Closest distance to the Earth

Planet	km	mi
Mercury	80 800 000	50 000 000
Venus	40 400 000	25 000 000
Mars	56 800 000	35 000 000
Jupiter	591 000 000	367 000 000
Saturn	1 198 000 000	744 000 000
Uranus	2 585 000 000	1 607 000 000
Pluto*	4 297 000 000	2 670 000 000
Neptune	4 308 000 000	2 678 000 000

*Between 1979 and 1999 Pluto will be closer to the Earth than Neptune because of the unusual shape of its orbit.

The solar system – Orbits and rotation

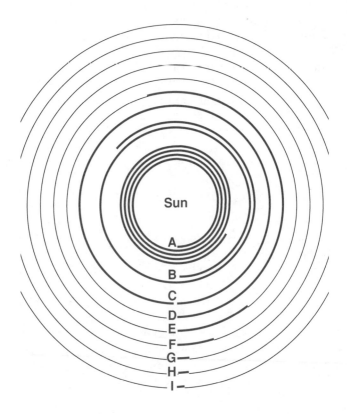

Sidereal period
Sidereal period is the time it takes a planet to orbit the
Sun. Planets' orbital speeds vary, as does their distance
from the Sun, so these periods are different for each
planet. The diagram shows how far each planet travels
in its orbit during the time it takes the Earth to
complete one orbit (approximately 1 year).

		Sidereal period	Average orbital speed
A	Mercury	88.0 days	47.9 km/s
B	Venus	224.7 days	35.0 km/s
C	Earth	365.3 days	29.8 km/s
D	Mars	687.0 days	24.1 km/s
E	Jupiter	11.86 years	13.1 km/s
F	Saturn	29.46 years	9.6 km/s
G	Uranus	84.01 years	6.8 km/s
H	Neptune	164.8 years	5.4 km/s
I	Pluto	247.7 years	4.7 km/s

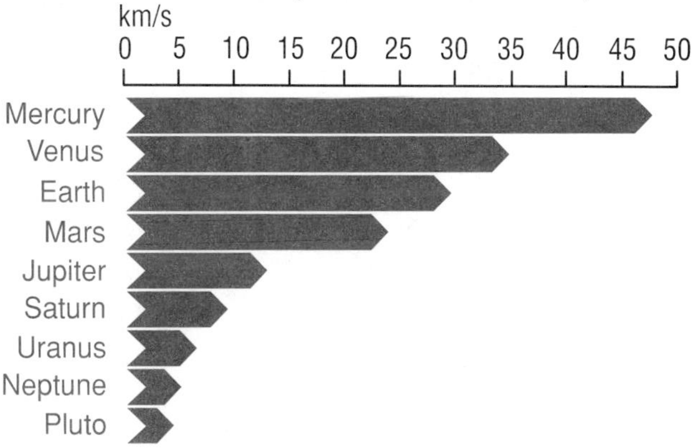

km/s
0   5   10   15   20   25   30   35   40   45   50

Mercury
Venus
Earth
Mars
Jupiter
Saturn
Uranus
Neptune
Pluto

Light years

The table below lists standard abbreviations and equivalents for the units used in measuring astronomical distances. These are very large units and are related to the Earth's orbit.

A light year (ly) is the distance light travels – at its speed of 299 792.458 km/s – through space over a tropical year.

An astronomical unit (au) is the mean distance between the Earth and the Sun.

A parsec (pc) is the distance at which a baseline of 1 au in length subtends an angle of 1 second.

1 au = 149 600 000 km = 93 000 000 mi
1 ly = 9 460 500 000 000 km = 5 878 000 000 000 mi
1 pc = 30 857 200 000 000 km = 19 174 000 000 000 mi
1 ly = 63 240 au
1 pc = 206 265 au = 3.262 ly

Planetary data

	Mercury	Venus	Earth
Mean distance from Sun	0.39 au	0.72 au	1.00 au
Distance at perihelion	0.31 au	0.72 au	0.98 au
Distance at aphelion	0.47 au	0.73 au	1.02 au
Closest distance to Earth	0.54 au	0.27 au	
Average orbital speed	47.9 km/s	35.0 km/s	29.8 km/s
Rotation period	58 days 15 hr	243 days	23 hr 56 min
Sidereal period	88 days	224.7 days	365.3 days
Diameter at equator	4878 km	12 104 km	12 756 km
Mass (Earth's mass=1)	0.06	0.82	1
Surface temperature	350 °C (day) −170 °C (night)	480 °C	22 °C
Gravity (Earth's gravity = 1)	0.38	0.88	1
Density (density of water = 1)	5.5	5.25	5.517
Number of satellites known	0	0	1
Number of rings known	0	0	0
Main gases in atmosphere	no atmosphere	Carbon dioxide	Nitrogen, oxygen

Planetary data (continued)

	Mars	Jupiter
Mean distance from Sun	1.52 au	5.20 au
Distance at perihelion	1.38 au	4.95 au
Distance at aphelion	1.67 au	5.46 au
Closest distance to Earth	0.38 au	3.95 au
Average orbital speed	24.1 km/s	13.1 km/s
Rotation period	24 hr 37 min	9 hr 50 min
Sidereal period	687 days	11.86 years
Diameter at equator	6795 km	142 800 km
Mass (Earth's mass=1)	0.11	317.9
Surface temperature	−23 °C	−150 °C
Gravity (Earth's gravity = 1)	0.38	2.64
Density (density of water = 1)	3.94	1.33
Number of satellites known	2	16
Number of rings known	0	1
Main gases in atmosphere	Carbon dioxide	Hydrogen, helium

Saturn	Uranus	Neptune	Pluto
9.54 au	19.18 au	30.06 au	39.36 au
9.01 au	18.28 au	29.80 au	29.58 au
10.07 au	20.09 au	30.32 au	49.14 au
8.01au	17.28 au	28.80 au	28.72 au
9.6 km/s	6.8 km/s	5.4 km/s	4.7 km/s
10 hr 14 min	16 hr 10 min	18 hr 26 min	6 days 9 hr
29.46 years	84.01 years	164.8 years	247.7 years
120 000 km	50 800 km	48 500 km	3000 km
95.2	14.6	17.2	0.002–0.003
−180 °C	−210 °C	−220 °C	−230 °C
1.15	1.17	1.2	not known
0.71	1.7	1.77	not known
19	5	2	1
1000+	9	0	0
Hydrogen, helium	Hydrogen, helium, methane	Hydrogen, helium, methane	Methane

12: Earth

Earth's interior

A Crust (under oceans) 6 km (4 mi) deep; made of basalt (a type of rock). Crust (continental): average 35 km (22 mi) deep; made of granite

B Mantle 2809 km (1745 mi) deep; probably containing peridotite (a heavy, dark rock), dunite (olivine rock) and ecologite (a dense form of basalt)

C Outer core 2000 km (1240 mi) deep; probably liquid iron with some dissolved sulphur and silicon

D Inner core 1482 km (920 mi) deep; probably solid iron

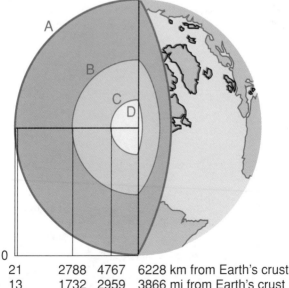

0				
21	2788	4767	6228	km from Earth's crust
13	1732	2959	3866	mi from Earth's crust

Atmospheric layers and depths of the Earth

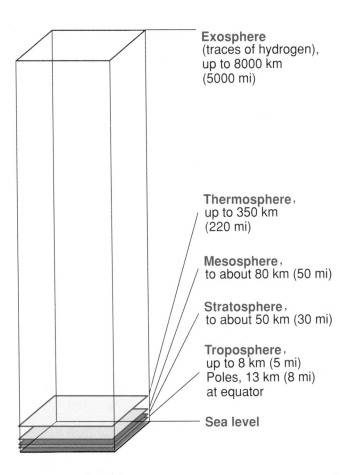

Exosphere
(traces of hydrogen),
up to 8000 km
(5000 mi)

Thermosphere,
up to 350 km
(220 mi)

Mesosphere,
to about 80 km (50 mi)

Stratosphere,
to about 50 km (30 mi)

Troposphere,
up to 8 km (5 mi)
Poles, 13 km (8 mi)
at equator

Sea level

World climates

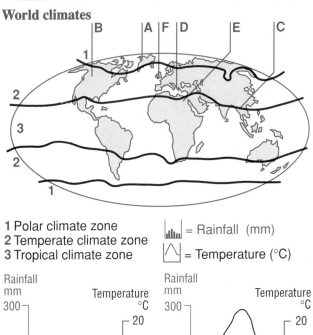

1 Polar climate zone
2 Temperate climate zone
3 Tropical climate zone

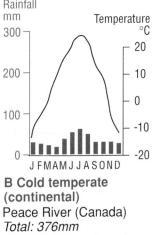

= Rainfall (mm)

= Temperature (°C)

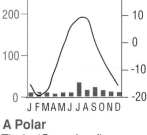

A Polar
Thule (Greenland)
Total: 93mm

**B Cold temperate
(continental)**
Peace River (Canada)
Total: 376mm

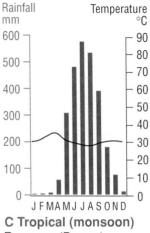

C Tropical (monsoon)
Rangoon (Burma)
Total: 2620mm

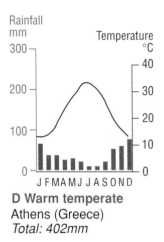

D Warm temperate
Athens (Greece)
Total: 402mm

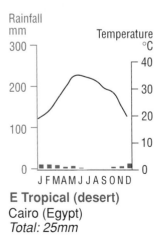

E Tropical (desert)
Cairo (Egypt)
Total: 25mm

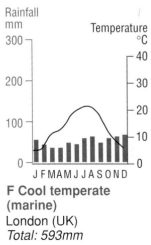

F Cool temperate (marine)
London (UK)
Total: 593mm

Continents

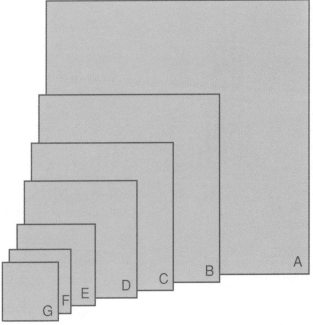

A Asia	44 250 000 km^2	17 085 000 mi^2
B Africa	30 264 000 km^2	11 685 000 mi^2
C N. America	24 398 000 km^2	9 420 000 mi^2
D S. America	17 793 000 km^2	6 870 000 mi^2
E Antarctica	13 209 000 km^2	5 100 000 mi^2
F Europe	9 907 000 km^2	3 825 000 mi^2
G Australia	8 534 000 km^2	3 295 000 mi^2

Largest countries

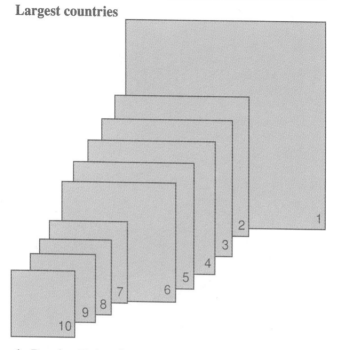

1	Russian Federation	17 075 000 km²	6 593 000 mi²
2	Canada	9 976 000 km²	3 852 000 mi²
3	China	9 561 000 km²	3 692 000 mi²
4	USA	9 520 000 km²	3 676 000 mi²
5	Brazil	8 512 000 km²	3 286 000 mi²
6	Australia	7 682 000 km²	2 966 000 mi²
7	India	3 288 000 km²	1 269 000 mi²
8	Argentina	2 777 000 km²	1 072 000 mi²
9	Sudan	2 506 000 km²	968 000 mi²
10	Zaïre	2 345 000 km²	905 000 mi²

Oceans and seas

		km^2	mi^2
1	Pacific Ocean	165 242 000	63 800 000
2	Atlantic Ocean	82 362 000	31 800 000
3	Indian Ocean	73 556 000	28 400 000
4	Arctic Ocean	13 986 000	5 400 000
5	South China Sea	2 975 000	1 149 000
6	Caribbean Sea	2 753 000	1 063 000
7	Mediterranean Sea	2 505 000	967 000
8	Bering Sea	2 269 000	876 000
9	Gulf of Mexico	1 544 000	596 000
10	Sea of Okhotsk	1 528 000	590 000

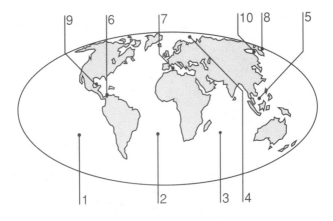

Largest islands

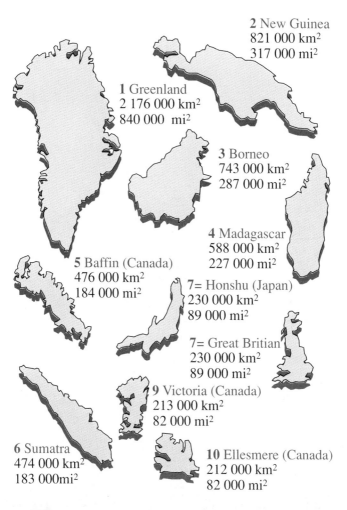

2 New Guinea
821 000 km²
317 000 mi²

1 Greenland
2 176 000 km²
840 000 mi²

3 Borneo
743 000 km²
287 000 mi²

4 Madagascar
588 000 km²
227 000 mi²

5 Baffin (Canada)
476 000 km²
184 000 mi²

7= Honshu (Japan)
230 000 km²
89 000 mi²

7= Great Britian
230 000 km²
89 000 mi²

9 Victoria (Canada)
213 000 km²
82 000 mi²

6 Sumatra
474 000 km²
183 000mi²

10 Ellesmere (Canada)
212 000 km²
82 000 mi²

Volcanoes and mountains

Highest volcanoes

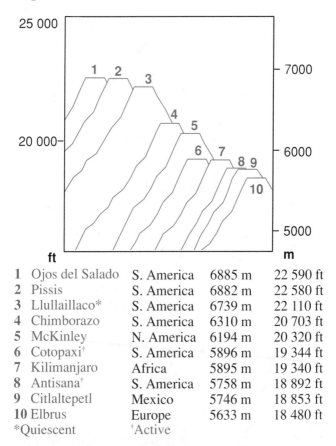

1	Ojos del Salado	S. America	6885 m	22 590 ft
2	Pissis	S. America	6882 m	22 580 ft
3	Llullaillaco*	S. America	6739 m	22 110 ft
4	Chimborazo	S. America	6310 m	20 703 ft
5	McKinley	N. America	6194 m	20 320 ft
6	Cotopaxi†	S. America	5896 m	19 344 ft
7	Kilimanjaro	Africa	5895 m	19 340 ft
8	Antisana†	S. America	5758 m	18 892 ft
9	Citlaltepetl	Mexico	5746 m	18 853 ft
10	Elbrus	Europe	5633 m	18 480 ft

*Quiescent   †Active

Highest mountains

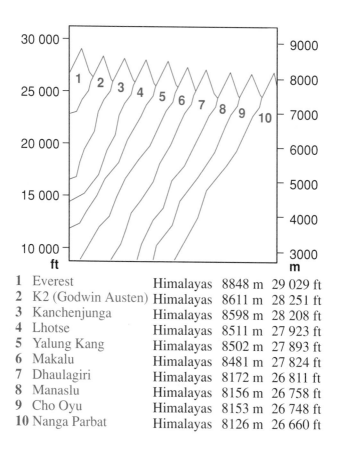

#	Name	Range	Height (m)	Height (ft)
1	Everest	Himalayas	8848 m	29 029 ft
2	K2 (Godwin Austen)	Himalayas	8611 m	28 251 ft
3	Kanchenjunga	Himalayas	8598 m	28 208 ft
4	Lhotse	Himalayas	8511 m	27 923 ft
5	Yalung Kang	Himalayas	8502 m	27 893 ft
6	Makalu	Himalayas	8481 m	27 824 ft
7	Dhaulagiri	Himalayas	8172 m	26 811 ft
8	Manaslu	Himalayas	8156 m	26 758 ft
9	Cho Oyu	Himalayas	8153 m	26 748 ft
10	Nanga Parbat	Himalayas	8126 m	26 660 ft

Highest mountain in each continent
When reading the figure, note the discontinuity
between 5000 ft and 15 000 ft, and 200 m and 4000 m.

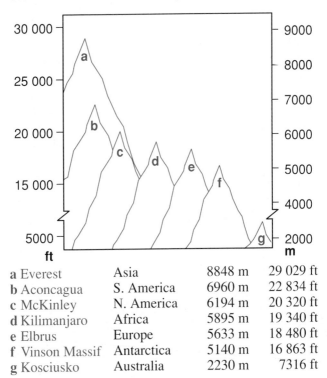

a Everest	Asia	8848 m	29 029 ft
b Aconcagua	S. America	6960 m	22 834 ft
c McKinley	N. America	6194 m	20 320 ft
d Kilimanjaro	Africa	5895 m	19 340 ft
e Elbrus	Europe	5633 m	18 480 ft
f Vinson Massif	Antarctica	5140 m	16 863 ft
g Kosciusko	Australia	2230 m	7316 ft

Highest mountain in selected countries

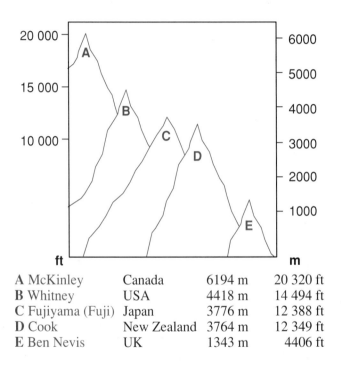

A McKinley	Canada	6194 m	20 320 ft
B Whitney	USA	4418 m	14 494 ft
C Fujiyama (Fuji)	Japan	3776 m	12 388 ft
D Cook	New Zealand	3764 m	12 349 ft
E Ben Nevis	UK	1343 m	4406 ft

Longest rivers

1	Nile	Africa	6650 km	4132 mi
2	Amazon	S. America	6437 km	4000 mi
3	Mississippi-Missouri-Red Rock	N. America	6212 km	3860 mi
4	Ob-Irtysh	Asia	5570 km	3461 mi
5	Yangtze (Chang)	Asia	5520 km	3430 mi
6	Huang He	Asia	4672 km	2903 mi
7	Congo (Zaire)	Africa	4667 km	2900 mi
8	Amur	Asia	4509 km	2802 mi
9	Lena	Asia	4270 km	2653 mi
10	Mackenzie	N. America	4241 km	2635 mi

Longest in its continent

A	Africa	Nile	6650 km	4132 mi
B	S. America	Amazon	6437 km	4000 mi
C	N. America	Mississippi-Missouri-Red Rock	6212 km	3860 mi
D	Asia	Ob-Irtysh	5570 km	3461 mi
E	Europe	Volga	3690 km	2293 mi
F	Australia	Murray	3219 km	2000 mi

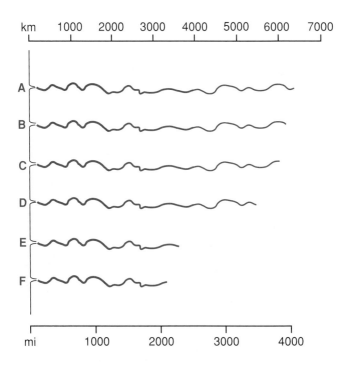

Largest lakes

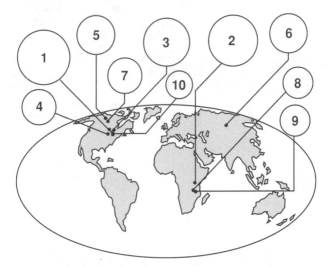

1	Superior	82 400 km^2	31 800 mi^2
2	Victoria	69 500 km^2	26 800 mi^2
3	Huron	59 600 km^2	23 000 mi^2
4	Michigan	58 000 km^2	22 400 mi^2
5	Great Bear	31 800 km^2	12 300 mi^2
6	Baykal	31 500 km^2	12 200 mi^2
7	Great Slave	28 400 km^2	11 000 mi^2
8	Tanganyika	28 400 km^2	11 000 mi^2
9	Malawi	28 200 km^2	10 900 mi^2
10	Erie	25 700 km^2	9 900 mi^2

Largest waterfalls

1 Angel, Venezuela
979 m (3212 ft)

2 Tugela, S. Africa
948 m (3110 ft)

3 Utigård, Norway
800 m (2625 ft)

4 Mongefossen, Norway
774 m (2540 ft)

5 Yosemite, USA
739 m (2425 ft)

6 Østre Mardøla Foss,
Norway
657 m (2154 ft)

7 Tyssestrengane, Norway
646 m (2120 ft)

8 Kukenaom, Venezuela
610 m (2000 ft)

9 Sutherland, N. Zealand
580 m (1904 ft)

10 Kjellfossen, Norway
561 m (1841 ft)

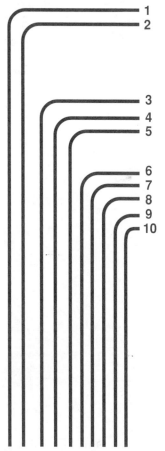

Largest deserts

A	Sahara	8 397 000 km²	3 242 000 mi²
B	Australian	1 549 000 km²	598 000 mi²
C	Arabian	1 300 000 km²	502 000 mi²
D	Gobi	1 039 000 km²	401 000 mi²
E	Kalahari	521 000 km²	201 000 mi²
F	Turkestan	360 000 km²	139 000 mi²
G	Takla Makan	321 000 km²	124 000 mi²
H	Sonoran	311 000 km²	120 000 mi²
I	Namib	311 000 km²	120 000 mi²
J	Thar	259 000 km²	100 000 mi²

K	United Kingdom	241 000 km²	93 000 mi²

The sizes of the largest deserts are compared (opposite) to the size of the United Kingdom.

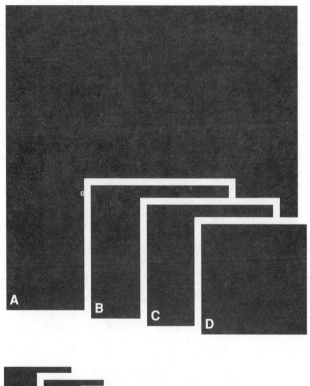

Deepest caves

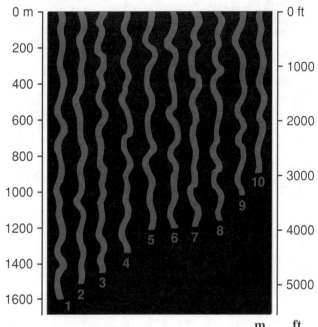

			m	ft
1	Réseau Jean Bernard	France	1602	5256
2	Shakta Pantjukhina	Russian Caucasus	1508	4947
3	Sistema del Trave	Spain	1441	4728
4	San Agustin	Mexico	1353	4439
5	Schwersystem	Austria	1219	3999
6	Abisso Olivifer	Italy	1210	3970
7	Veliko Fbrego	Former Yugoslavia	1198	3930
8	Anou Ifflis	Algeria	1159	3802
9	Siebenhengste System	Switzerland	1020	3346
10	Jama u Vjetrena brda	Former Yugoslavia	897	2943

Capitals of the world
Africa
ALGERIA Algiers
ANGOLA Luanda
BENIN Porto-Novo
BOTSWANA Gaborone
BURKINA FASO
  Ouagadougou
BURUNDI Bujumbura
CAMEROON Yaoundé
CAPE VERDE Praia
CENTRAL AFRICAN
REPUBLIC Bangui
CHAD N'Djamena
COMOROS Moroni
CONGO Brazzaville
DJIBOUTI Djibouti
EGYPT Cairo
EQUATORIAL GUINEA
  Malabo
ERITREA Asmara
ETHIOPIA Addis Ababa
GABON Libreville
GAMBIA Banjul
GHANA Accra
GUINEA Conakry
GUINEA-BISSAU
  Bissau
IVORY COAST
(CÔTE D'IVOIRE)
  Yamoussoukro/Abidjan
KENYA Nairobi

LESOTHO Maseru
LIBERIA Monrovia
LIBYA Tripoli
MADAGASCAR
  Antananarivo
MALAWI Lilongwe
MALI Bamako
MAURITANIA
  Nouakchott
MAURITIUS Port Louis
MOROCCO Rabat
MOZAMBIQUE Maputo
NAMIBIA Windhoek
NIGER Niamey
NIGERIA Abuja
RWANDA Kigali
SÃO TOMÉ AND
PRÍNCIPE São Tomé
SENEGAL Dakar
SEYCHELLES Victoria
SIERRA LEONE Freetown
SOMALIA Mogadishu
SOUTH AFRICA
  Cape Town/ Pretoria
SUDAN Khartoum
SWAZILAND Mbabane
TANZANIA Dodoma
TOGO Lomé
TUNISIA Tunis
UGANDA Kampala
ZAÏRE Kinshasa
ZAMBIA Lusaka

ZIMBABWE Harare

Asia and Middle East
AFGHANISTAN Kabul
BAHRAIN Manama
BANGLADESH Dhaka
BHUTAN Thimphu
BRUNEI Bandar Seri
  Begawan
CAMBODIA
  Phnom Penh
CHINA Beijing
INDIA New Delhi
INDONESIA Jakarta
IRAN Tehran
IRAQ Baghdad
ISRAEL Jerusalem
JAPAN Tokyo
JORDAN Amman
KAZAKHSTAN Alma-
  Ata
KIRGHIZIA Frunze
KOREA, NORTH
  Pyongyang
KOREA, SOUTH Seoul
KUWAIT Kuwait City
LAOS Vientiane
LEBANON Beirut
MALAYSIA
  Kuala Lumpur
MALDIVES Malé
MONGOLIA Ulan Bator

MYANMAR(Burma)
  Yangon (Rangoon)
NEPAL Kathmandu
OMAN Muscat
PAKISTAN Islamabad
PHILIPPINES Manila
QATAR Doha
SAUDI ARABIA Riyadh
SINGAPORE Singapore
SRI LANKA Colombo
SYRIA Damascus
TADZHIKISTAN
  Dushanbe
THAILAND Bangkok
TURKMENISTAN
  Ashkhabad
UNITED ARAB
  EMIRATES Abu Dhabi
UZBEKISTAN Tashkent
VIETNAM Hanoi
YEMEN Sana'a

Europe
ALBANIA Tirana
ANDORRA
  Andorra la Vella
ARMENIA Yerevan
AUSTRIA Vienna
AZERBAIJAN Baku
BELARUS Minsk
BELGIUM Brussels
BULGARIA Sofia

CROATIA Zagreb
CYPRUS Nicosia
CZECHOSLOVAKIA
   Prague
DENMARK Copenhagen
ESTONIA Tallinn
FINLAND Helsinki
FRANCE Paris
GEORGIA Tbilisi
GERMANY Berlin
GREECE Athens
HUNGARY Budapest
ICELAND Reykjavík
IRELAND (Eire) Dublin
ITALY Rome
LATVIA Riga
LIECHTENSTEIN Vaduz
LITHUANIA Vilnius
LUXEMBOURG
   Luxembourg
MALTA Valletta
MOLDOVA Kishinev
MONACO Monaco-Ville
NETHERLANDS
   The Hague/Amsterdam
NORWAY Oslo
POLAND Warsaw
PORTUGAL Lisbon
ROMANIA Bucharest
RUSSIA Moscow
SAN MARINO
   San Marino

SLOVENIA Ljubljana
SPAIN Madrid
SWEDEN Stockholm
SWITZERLAND Bern
TURKEY Ankara
UKRAINE Kiev
UNITED KINGDOM
   London
VATICAN CITY
   Vatican city
YUGOSLAVIA Belgrade
   (in 1991; in 1992 not
   known)

Australasia
AUSTRALIA Canberra
FIJI Suva
KIRIBATI Tarawa
MARSHALL ISLANDS
   Dalap-Uliga-Darrit
MICRONESIA
   Kolonia
NAURU Yaren
NEW ZEALAND
   Wellington
PALAU Koror
PAPUA NEW GUINEA
   Port Moresby
SOLOMON ISLANDS
   Honiara
TONGA Nuku'alofa
TUVALU Funafuti

VANUATU Port-Vila
WESTERN SAMOA Apia

South America
ARGENTINA Buenos
  Aires
BOLIVIA La Paz/Sucre
BRAZIL Brasília
CHILE Santiago
COLOMBIA Bogotá
ECUADOR Quito
FRENCH GUYANA
  Cayenne
GUYANA Georgetown
PARAGUAY Asunción
PERU Lima
SURINAME Paramaribo
URUGUAY Montevideo
VENEZUELA Caracas

**North and Central
America**
ANTIGUA AND
  BARBUDA St John's
BAHAMAS Nassau
BARBADOS Bridgetown
BELIZE Belmopan
CANADA Ottawa
COSTA RICA San José

CUBA Havana
DOMINICA Roseau
DOMINICAN
REPUBLIC Santo
  Domingo
EL SALVADOR
  San Salvador
GREENLAND Nuuk
GRENADA St George's
GUATEMALA
  Guatemala City
HAITI Port-au-Prince
HONDURAS
  Tegucigalpa
JAMAICA Kingston
MEXICO Mexico City
NICARAGUA Managua
PANAMA Panama City
ST LUCIA Castries
ST CHRISTOPHER AND
  NEVIS Basseterre
ST VINCENT AND THE
  GRENADINES
  Kingstown
TRINIDAD AND
  TOBAGO Port of Spain
UNITED STATES OF
  AMERICA
  Washington DC

Distances of world cities

Listed here are the approximate distances of principal airports from London.

Name	Distance (mi)	(km)	Name	Distance (mi)	(km)
Athens	1 500	2 400	Mexico City	5 500	9 000
Bahrain	3 200	5 100	Montreal	3 200	5 200
Bangkok	6 000	9 500	Moscow	1 600	2 500
Bombay	4 500	7 200	Nairobi	4 200	6 800
Buenos Aires	7 000	11 100	New York	3 400	5 500
Cairo	2 200	3 500	Paris	200	400
Chicago	4 000	6 300	Peking	5000	8 100
Frankfurt	400	700	Rio de Janeiro	5 800	9 200
Hong Kong	6 000	9 700	Rome	900	1 500
Johannesburg	5 700	9 100	San Francisco	5 300	8 700
Karachi	4 000	6 300	Singapore	6 800	10 900
Lagos	3 200	5 000	Sydney	10 600	17 000
Lima	6 300	10 100	Teheran	2 800	4 400
Madrid	800	1 200	Tokyo	6 000	9 600
Manila	6 700	10 800	Vancouver	4 800	7 600

Currencies of the world

Currency	Country where used
Afghani	Afghanistan
Austral	Argentina
Baht	Thailand
Balboa	Panama
Bipkwele	Equatorial Guinea
Birr	Ethiopia
Bolivar	Venezuela
Boliviano	Bolivia
Cedi	Ghana
Colón	Costa Rica, El Salvador
Córdoba	Nicaragua
Cruzeiro	Brazil
Dalasi	The Gambia
Dinar	Algeria, Bahrain, Iraq, Jordan, Kuwait, Libya, Tunisia
Dirham	Morocco, United Arab Emirates
Dobra	São Tomé and Príncipe
Dollar	Antigua and Barbuda, Australia, the Bahamas, Barbados, Belize, Brunei Darussalam, Canada, Dominica, Fiji, Grenada, Guyana, Hong Kong, Jamaica, Kiribati, Liberia, Nauru, New Zealand, St. Christopher and Nevis, St. Lucia, St. Vincent and the Grenadines, Singapore, Solomon Is, Taiwan, Trinidad and Tobago, Tuvalu, USA, Vanuatu, Zimbabwe
Dong	Vietnam
Drachma	Greece
Duktat	Belarus
Escudo	Cape Verde, Portugal

Currency	Country where used	Currency	Country where used
Forint	Hungary	Krona	Sweden
Franc	Andorra, Belgium, Benin, Burkina Faso, Burundi, Cameroon, Central African Republic, Chad, Comoros, Congo, Djibouti, France, Gabon, Guinea, Ivory Coast, Liechtenstein, Luxembourg, Madagascar, Mali, Monaco, Niger, Rwanda, Senegal, Switzerland, Togo, Vanuatu	Krone	Denmark
		Kroner	Norway
		Kronur	Iceland
		Kroon	Estonia
		Kwacha	Malawi, Zambia
		Kwanza	Angola
		Kyat	Myanmar
		Lei	Romania
		Lek	Albania
		Lempira	Honduras
		Leone	Sierra Leone
		Leu	Romania
		Lev	Bulgaria
		Lilangeni	Swaziland
		Lira	Italy, San Marino, Turkey, Vatican City
Gourde	Haiti	Maloti	Lesotho
Guarani	Paraguay	Manat	Azerbaijan
Guilder	Netherlands, Suriname	Mark	Germany
		Markka	Finland
		Metical	Mozambique
Hryvnia	Ukraine	Naira	Nigeria
Kina	Papua New Guinea	New Dinar	Former Yugoslavia
Koruna	Czech Republic, Slovak Republic	New Kip	Laos
		New Kwanza	Angola
Krona	Iceland	New Peso	Uruguay

Currency	Country where used	Currency	Country where used
New sol	Peru	Riel	Cambodia
Ngultrum	Bhutan	Ringgit	Malaysia
Ouguija	Mauritania	Riyal	Qatar, Saudi Arabia, Yemen
Pa'anga	Tonga		
Pataca	Macao	Rouble	Former USSR
Peseta	Andorra, Spain	Rufiyaa	Maldives
Peso	Bolivia, Chile, Cuba, Colombia, Dominican Republic, Guinea-Bissau, Mexico, Philippines	Rupee	India, Mauritius, Nepal, Pakistan, Seychelles, Sri Lanka
		Rupiah	Indonesia
		Schilling	Austria
		Shekel	Israel
Pound	Cyprus, Egypt, Ireland, Lebanon, Malta, Sudan, Syria, United Kingdom and Northern Ireland	Shilling	Kenya, Somalia, Tanzania, Uganda
		Soum	Uzbekistan
		Sucre	Ecuador
		Taka	Bangladesh
		Tala	Western Samoa
Pula	Botswana	Tugrik	Mongolia
Punt	Republic of Ireland	Won	North Korea, South Korea
Quetzal	Guatemala	Yen	Japan
Rand	Namibia, South Africa	Yuan	China
		Zaire	Zaire
Rial	Iran, Oman	Zloty	Poland

13: Symbols, codes and alphabets

International dialling codes
For calls to any of the countries listed here first
dial 010

Afganistan	93	Benin	229
Albania	355	Bermuda	1 809
Algeria	213	Bhutan	975
Andorra	33 628	Bolivia	591
Angola	244	Bosnia-Herzegovina	387
Anguilla	1 809	Botswana	267
Antigua and		Brazil	55
Barbuda	1 809	Brunei Darussalam	673
Antilles	599	Bulgaria	359
Argentina	54	Burkina Faso	226
Aruba	2 978	Burundi	257
Ascension Island	247	Cameroon	237
Australia	61	Canada	1
Austria	43	Canary Island	34
Azores	351	Cape Verde Islands	238
Bahamas	1 809	Cayman Islands	1 809
Bahrain	973	Central African	
Bangladesh	880	Republic	236
Barbados	1 809	Chad	235
Belgium	32	Chile	56
Belize	501	China	86

Christmas Island	6 724	Gabon	241
CIS	7	Gambia	220
Cocos Island	6 722	Germany	49
Colombia	57	Ghana	233
Comoros	269	Gibraltar	350
Congo	242	Greece	30
Cook Islands	682	Greenland	299
Costa Rica	506	Grenada	1 809
Côte d'Ivoire	225	Guadeloupe	590
Croatia	385	Guam	671
Cuba	53	Guatemala	502
Cyprus	357	Guinea	224
Czech Republic	42	Guinea-Bissau	245
Denmark	45	Guyana	592
Djibouti	253	Haiti	509
Dominica	1 809	Honduras	504
Dominican Republic	1809	Hong Kong	852
Ecuador	593	Hungary	36
Egypt	20	Iceland	354
El Salvador	503	India	91
Equatorial Guinea	240	Indonesia	62
Estonia	7	Iran	98
Ethiopia	251	Iraq	964
Falkland Islands	500	Ireland	353
Faroe Islands	298	Israel and the	
Fiji	679	Occupied Territories	972
Finland	358	Italy	39
France	33	Jamaica	1 809
French Polynesia	689	Japan	81

Jordan	962	Micronesia	691
Kenya	254	Monaco	33 93
Kiribatt	686	Montserrat	1 809
Korea North	850	Morocco	212
Korea South	82	Mozambique	258
Kuwait	965	Myanmar	95
Latvia	7	Namibia	264
Lebanon	961	Nauru	674
Lesotho	266	Nepal	977
Liberia	231	Netherlands	31
Libya	218	New Caledonia	687
Liechtenstein	41 75	New Zealand	64
Lithuania	7	Nicaragua	505
Luxembourg	352	Niger	227
Macao	853	Nigeria	234
Macedonia	389	Niue	683
Madagascar	261	Norfolk Island	6 723
Madeira	351 91	Northern Marianas	670
Malawi	265	Norway	47
Malaysia	60	Oman	968
Maldives	960	Pakistan	92
Mali	223	Palau	6809
Malta	356	Panama	507
Marshall Islands	692	Papua New Guinea	675
Martinique	596	Paraguay	595
Mauritania	222	Peru	51
Mauritius	230	Phillipines	63
Mayotte	269	Poland	48
Mexico	52	Portugal	351

Puerto Rico	1 809	Sudan	249
Qatar	974	Suriname	597
Reunion	262	Swaziland	268
Romania	40	Sweden	46
Rwanda	250	Switzerland	41
St Helena	290	Syria	963
St Kitts and Nevis	1 809	Taiwan	886
St Lucia	1 809	Tanzania	255
St Pierre and Miquelon	508	Thailand	66
St Vincent and		Togo	228
the Grenadines	1 809	Tonga	676
Samoa USA	684	Trinidad and Tobago	1809
Somoa Western	685	Tunisia	216
San Marino	39 549	Turkey	90
São Tomé,		Turks, Caicos	1 809
Príncipe	23 912	Tuvalu	688
Saudi Arabia	966	Uganda	256
Senegal	221	United Arab Emirates	971
Serbia	381	Uruguay	598
Seychelles	248	USA	1
Sierra Leone	232	Vanuatu	678
Singapore	65	Venezuela	58
Slovak Republic	42	Vietnam	84
Slovenia	386	Virgin Isl (UK)	1 809 49
Solomon Islands	677	Virgin Isl (US)	1 809
Somalia	252	Yemen	967
South Africa	27	Zaire	243
Spain	34	Zambia	260
Sri Lanka	94	Zimbabwe	263

Car registration letters

* Countries where driving on the left prevails

A	Austria	DK	Denmark
ADN	Yemen PDR	DOM	Dominican Republic
AFG	Afghanistan		
AL	Albania	DY	Benin
AND	Andorra	DZ	Algeria
AUS	Australia*	E	Spain
B	Belgium	EAK	Kenya*
BD	Bangladesh*	EAZ	Tanzania*
BDS	Barbados*	EAU	Uganda*
BG	Bulgaria	EC	Ecuador
BH	Belize	ES	El Salvador
BR	Brazil	ET	Egypt
BRN	Bahrain*	ETH	Ethiopia
BRU	Brunei*	F	France
BS	Bahamas*	FJI	Fiji*
BUR	Burma*	FL	Liechtenstein
C	Cuba*	FR	Faroe Is
CDN	Canada	GB	Great Britain*
CH	Switzerland	GBA	Alderney*
CI	Côte d'Ivoire	GBG	Guernsey*
CL	Sri Lanka*	GBJ	Jersey*
CO	Colombia	GBM	Isle of Man*
CR	Costa Rica	GBZ	Gibraltar
CZ	Czechoslovakia	GCA	Guatemala
CY	Cyprus*	GH	Ghana
D	Germany	GR	Greece

GUY	Guyana*
H	Hungary
HK	Hong Kong*
HJK	Jordan
I	Italy
IL	Israel
IND	India*
IR	Iran
IRL	Ireland*
IRQ	Iraq
IS	Iceland*
J	Japan
JA	Jamaica*
K	Kampuchea
KWT	Kuwait
L	Luxembourg
LAO	Laos PDR
LAR	Libya
LB	Liberia
LS	Lesotho*
MA	Morocco
MAL	Malaysia*
MC	Monaco
MEX	Mexico
MS	Mauritius*
MW	Malawi
N	Norway
NA	Netherlands Antilles

NIC	Nicaragua
NL	Netherlands
NZ	New Zealand*
P	Portugal
PA	Panama
PAK	Pakistan*
PE	Peru
PL	Poland
PNG	Papua New Guinea*
PY	Paraguay*
R	Romania
RA	Argentina
RB	Botswana*
RC	Taiwan
RCA	Central African Republic
RCB	Congo
RCH	Chile
RH	Haiti
RI	Indonesia
RIM	Mauritania
RL	Lebanon
RM	Madagascar
RMM	Mali
RN	Niger
ROK	Republic of South Korea
RP	Philippines

RSM	San Marino
RU	Burundi
RWA	Rwanda
S	Sweden*
SD	Swaziland*
SF	Finland
SGP	Singapore*
SME	Suriname
SN	Senegal
SU	CIS
SWA	Namibia
SY	Seychelles
SYR	Syria
T	Thailand*
TG	Togo
TN	Tunisia
TR	Turkey
TD	Trinidad and Tobago*
U	Uruguay
USA	USA
V	Vatican City
VN	Vietnam
WAG	Gambia*
WAL	Sierra Leone*
WAN	Nigeria*
WD	Dominica*
WG	Grenada*
WL	St Lucia*
WS	West Samoa
WV	St Vincent and the Grenadines*
YV	Venezuela
Z	Zambia*
ZA	South Africa*
ZRE	Zaire
ZW	Zimbabwe*

Dewey Decimal library system

000-099 GENERAL WORKS
almanacs, encyclopedias, bibliographies,
magazines, newspapers, materials that cannot
be narrowed to a single subject

100-199 PHILOSOPHY
logic, history of philosophy, systems of
philosophy, ethics, psychology

200-299 RELIGION
sacred writings (the Bible), mythology
history of religions, all religions and theologies

300-399 SOCIOLOGY (social sciences)
group dynamics, government, education,
economics

400-499 LANGUAGE (study of linguistics)
dictionaries dealing with words (not of
biographies), grammar and technical studies
of all languages

500-599 SCIENCE (Subject and theoretical)
Astronomy, biology, botany, chemistry
mathematics, physics

600-699 APPLIED SCIENCE (useful arts)
Agriculture, all types of engineering, business
home economics, medicine, nursing

700-799 ARTS (Professional and recreative)
architecture, painting, music, performing arts
sports

800-899 LITERATURE
All types of literature—drama, essays, novels, poetry—in all languages of all countries

900-999 HISTORY
all history, biography, geography and travel

Bar codes
Bar codes are a method of labelling items in computer-readable form for purposes of stock control, repeat ordering, confirmation of price, and so on.

The bar code is a set of binary numbers, represented as a pattern of black and white vertical lines of different thicknesses (1), with a series of numbers above or below, and sometimes both (2). The lines are a machine-readable representation of the numbers. When the bar-code is passed across an optical scanner, the information is sent to a central computer, where it is recorded on the computer's hard disc.

Bar codes are most widely used in the retail trade, and are a familiar feature of supermarket shopping. They are also used in libraries to control book circulation, and in factories to ensure that the correct parts are assembled.

Bar codes are subject to regulatory controls such as the European Article Numbering Code and the Universal Product Code in the US.

Hebrew

Russian

a	b	v	g	d
Аа	Бб	Вв	Гг	Дд
e	zh	z	i	
Ее	Жж	Зз	Ии	
k	l	m	n	
Кк	Лл	Мм	Нн	
o	p	r	s	t
Оо	Пп	Рр	Сс	Тт
u	f	kh	ts	ch
Уу	Фф	Хх	Цц	Чч
sh	shch	(hard sign)	y	
Шш	Щщ	Ъъ	Ыы	
(soft sign)	e	yu	ya	
Ьь	Ээ	Юю	Яя	

Greek

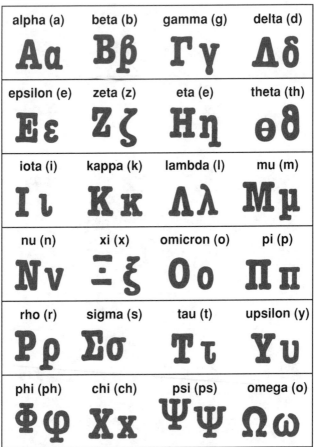

alpha (a)	beta (b)	gamma (g)	delta (d)
Α α	Β β	Γ γ	Δ δ
epsilon (e)	zeta (z)	eta (e)	theta (th)
Ε ε	Ζ ζ	Η η	Θ θ
iota (i)	kappa (k)	lambda (l)	mu (m)
Ι ι	Κ κ	Λ λ	Μ μ
nu (n)	xi (x)	omicron (o)	pi (p)
Ν ν	Ξ ξ	Ο ο	Π π
rho (r)	sigma (s)	tau (t)	upsilon (y)
Ρ ρ	Σ σ	Τ τ	Υ υ
phi (ph)	chi (ch)	psi (ps)	omega (o)
Φ φ	Χ χ	Ψ ψ	Ω ω

Braille

A	B	C	D	E	F

G	H	I	J	K	L

M	N	O	P	Q	R

S	T	U	V	W	X

Y	Z	1	2	3	4

5	6	7	8	9	0

Morse code

A	B	C	D
● ▬	▬ ● ● ●	▬ ● ▬ ●	▬ ● ●
E	**F**	**G**	**H**
●	● ● ▬ ●	▬ ▬ ●	● ● ● ●
I	**J**	**K**	**L**
● ●	● ▬ ▬ ▬	▬ ● ▬	● ▬ ● ●
M	**N**	**O**	**P**
▬ ▬	▬ ●	▬ ▬ ▬	● ▬ ▬ ●
Q	**R**	**S**	**T**
▬ ▬ ● ▬	● ▬ ●	● ● ●	▬
U	**V**	**W**	**X**
● ● ▬	● ● ● ▬	● ▬ ▬	▬ ● ● ▬
Y	**Z**	**1**	**2**
▬ ● ▬ ▬	▬ ▬ ● ●	● ▬ ▬ ▬ ▬	● ● ▬ ▬ ▬
3	**4**	**5**	**6**
● ● ● ▬ ▬	● ● ● ● ▬	● ● ● ● ●	▬ ● ● ● ●
7	**8**	**9**	**0**
▬ ▬ ● ● ●	▬ ▬ ▬ ● ●	▬ ▬ ▬ ▬ ●	▬ ▬ ▬ ▬ ▬
Comma	**Question mark**	**Error**	**Semicolon**
▬ ▬ ● ● ▬ ▬	● ● ▬ ▬ ● ●	● ● ● ● ● ● ● ●	▬ ● ▬ ● ▬ ●
Quotation marks	**Understand**	**Colon**	**Wait**
● ▬ ● ● ▬ ●	● ▬ ●	▬ ▬ ▬ ● ● ●	● ▬ ● ● ●
Hyphen	**Period**	**End of message**	**Apostrophe**
▬ ● ● ● ● ▬	● ▬ ● ▬ ● ▬	● ▬ ● ▬ ●	● ▬ ▬ ▬ ▬ ●

Semaphore

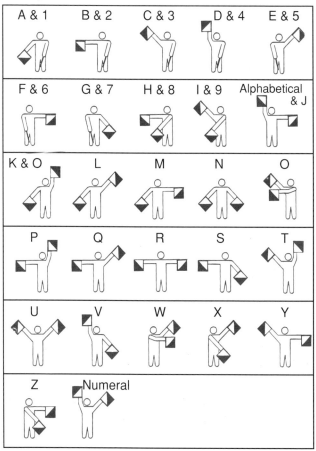

A & 1 B & 2 C & 3 D & 4 E & 5

F & 6 G & 7 H & 8 I & 9 Alphabetical & J

K & O L M N O

P Q R S T

U V W X Y

Z Numeral

Manual alphabet *(mostly used by deaf mutes)*

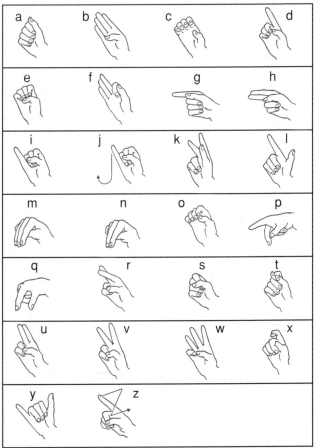

Manual numerals *(mostly used by deaf mutes)*

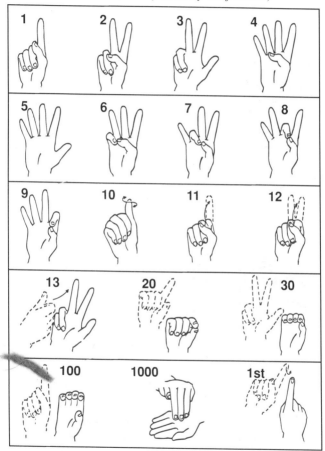

Flag signals at sea

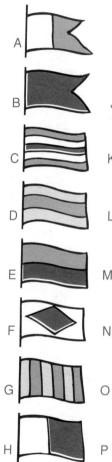

A I am undergoing speed trials

B I have explosives on board

C Yes

D Keep clear, I am in difficulty

E I am altering course to starboard

F I am disabled

G I require a pilot

H Pilot is on board

I I am altering course to port

J I am sending a message by semaphore

K Stop at once

L Stop, I wish to communicate with you

M A doctor is on board

N No

O Man overboard

P (The Blue Peter) I am about to sail

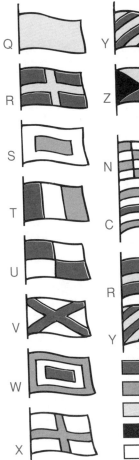

Red — Red
Blue — Blue
Yellow — Yellow
Black — Black
White — White

Q Quarantine flag
R I have stopped
S I am going astern
T Do not pass ahead of me
U You are in danger
V l need help
W Send a doctor
X Stop, and watch for my signals
Y I am carrying mail
Z I am calling a station ashore

N & C flying together "In distress: need inmediate assistance"
R & Y flying together from masthead "Crew has mutinied"

Musical notation 1

Musical notation 2

Astronomical notation

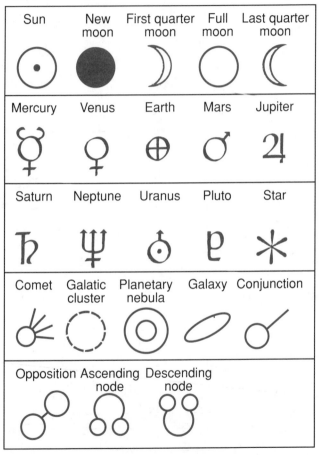

Sun	New moon	First quarter moon	Full moon	Last quarter moon
Mercury	Venus	Earth	Mars	Jupiter
Saturn	Neptune	Uranus	Pluto	Star
Comet	Galatic cluster	Planetary nebula	Galaxy	Conjunction
Opposition	Ascending node	Descending node		

Religious symbols

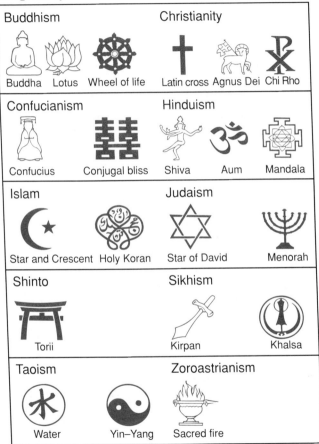

Buddhism

Buddha Lotus Wheel of life

Christianity

Latin cross Agnus Dei Chi Rho

Confucianism

Confucius Conjugal bliss

Hinduism

Shiva Aum Mandala

Islam

Star and Crescent Holy Koran

Judaism

Star of David Menorah

Shinto

Torii

Sikhism

Kirpan Khalsa

Taoism

Water Yin–Yang

Zoroastrianism

Sacred fire

Proofreader's marks

Margin mark	Text mark	Meaning
STET	the under copy	leave unchanged
⊗	the circle round copy	wrong font
New text before ⋏	⋏ between words	insert copy
ꝗ	the through copy	delete
⏟	the under copy	set in italics
⌇	the under copy	set in bold
CAPS	the under characters	set in capitals
⌇	the under copy	set in bold italics
≠	The circle round copy	set in lower case
⊙	⋏ where required	insert full stop
⸲/⊙/⸴/	⋏ where required	insert comma, colon semi-colon etc.
N.P.	⌐ between words	new paragraph
	⌐ between paragraphs	run on
⋎#	e ⋏a between characters	insert space
⋎ #	as⋏s between words	insert space
⋏ #	as ⋏is between characters	reduce space
⋏ #	e ⋏a between words	reduce space
⊢⊣	⋏ where required	insert hyphen
⊢⊣ 4mm	⋏ where required	insert rule (specify rule size in margin)
#	e ⋏a between characters	equalize spacing

Hazard warning symbols
Shown here are hazard warning symbols commonly displayed on commercial vehicles, trains, industrial sites and commercial products.

Corrosive

Explosive

High voltage

Highly
flammable

Oxidizing

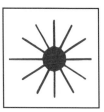

Radiation
(laser)

Radiation
(non-ionising)

Radioactive

Toxic

14: Abbreviations

Counties of England

Common abbreviation	Official name	Administrative headquarters
Avon	Avon	Bristol
Beds	Bedfordshire	Bedford
Berks	Berkshire	Reading
Bucks	Buckinghamshire	Aylesbury
Cambs	Cambridgeshire	Cambridge
–	Cheshire	Chester
–	Cleveland	Middlesborough
–	Cornwall	Truro
–	Cumbria	Carlisle
–	Derbyshire	Matlock
–	Devon	Exeter
–	Dorset	Dorchester
Co. Durham	County Durham	Durham
E. Sussex	East Sussex	Lewes
–	Essex	Chelmsford
Glos	Gloucestershire	Gloucester
–	Greater London	
–	Greater Manchester	
Hants	Hampshire	Winchester
–	Hereford and Worcester	Worcester
Herts	Hertfordshire	Hertford
–	Humberside	Beverley
I.O.W.	Isle of Wight	Newport
–	Kent	Maidstone

These pages give commonly-used abbreviations, grouped together by kind. The section lists geographical abbreviations, UN and other international organizations, languages, Christianity, science, units of time, speed, medicine and everyday abbreviations.

Common abbreviation	Official name	Administrative headquarters
Lancs	Lancashire	Preston
Leics	Leicestershire	Leicester
Lincs	Lincolnshire	Lincoln
–	Merseyside	
–	Norfolk	Norwich
Northants	Northamptonshire	Northampton
–	Northumberland	Morpeth
N. Yorks	North Yorkshire	Northallerton
Notts	Nottinghamshire	Nottingham
Oxon	Oxfordshire	Oxford
Salop	Shropshire	Shrewsbury
–	Somerset	Taunton
S. Yorks	South Yorkshire	
Staffs	Staffordshire	Stafford
–	Suffolk	Ipswich
–	Surrey	Kingston upon Thames
–	Tyne and Wear	
Warwks	Warwickshire	Warwick
W. Midlands	West Midlands	
W. Sussex	West Sussex	Chichester
W. Yorks	West Yorkshire	
Wilts	Wiltshire	Trowbridge

Counties of Wales

Abbreviation	County	Administrative headquarters
–	Clwyd	Mold
–	Dyfed	Carmarthen
–	Gwent	Cwmbran
–	Gwynedd	Caernarfon
M. Glam	Mid Glamorgan	Cardiff
–	Powys	Llandrindod Wells
S. Glam	South Glamorgan	Cardiff
W. Glam	West Glamorgan	Swansea

Regions of Scotland

Abbreviation	Region	Administrative headquarters
–	Borders	Newtown St. Boswells
–	Central	Stirling
–	Dumfries and Galloway	Dumfries
–	Fife	Glenrothes
–	Grampian	Aberdeen
–	Highland	Inverness
–	Lothian	Edinburgh
–	Orkney Islands	Kirkwall
–	Shetland Islands	Lerwick
–	Strathclyde	Glasgow
–	Tayside	Dundee
–	Western Isles	Stornoway

Counties of Ireland
Northern Ireland

Abbreviation	County	Administrative headquarters
Ant	Antrim	Antrim
Arm	Armagh	Armagh
–	Down	Downpatrick
Ferm	Fermanagh	Enniskillen
Lond/co Derry	Londonderry	Londonderry
Tyr	Tyrone	Omagh

Republic of Ireland

Abb.	County	Abb.	County
Car	Carlow	Long	Longford
Cav	Cavan	–	Louth
–	Clare	–	Mayo
–	Cork	–	Meath
Don	Donegal	Monag	Monaghan
Dub	Dublin	Off	Offaly
Gal	Galway	Ros	Roscommon
Ker	Kerry	–	Sligo
Kild	Kildare	Tipp	Tipperary
Kilk	Kilkenny	Wat	Waterford
Leit	Leitrim	–	Westmeath
–	Laois	Wex	Wexford
Lim	Limerick	Wick	Wicklow

States of the United States of America

Abbreviation	Zip code abbreviation	State	State capital
Ala.	AL	Alabama	Montgomery
Alas.	AK	Alaska	Juneau
Ariz.	AZ	Arizona	Phoenix
Ark.	AR	Arkansas	Little Rock
Calif.	CA	California	Sacramento
Col.	CO	Colorado	Denver
Conn.	CT	Connecticut	Hartford
Del.	DE	Delaware	Dover
Fla.	FL	Florida	Tallahassee
Ga.	GA	Georgia	Atlanta
–	HI	Hawaii	Honolulu
Ida.	ID	Idaho	Boise
Ill.	IL	Illinois	Springfield
Ind.	IN	Indiana	Indianapolis
Ia.	IA	Iowa	Des Moines
Kan.	KS	Kansas	Topeka
Ky.	KY	Kentucky	Frankfort
La.	LA	Louisiana	Baton Rouge
Me.	ME	Maine	Augusta
Md.	MD	Maryland	Annapolis
Mass.	MA	Massachusetts	Boston
Mich.	MI	Michigan	Lansing
Minn.	MN	Minnesota	St Paul
Miss.	MS	Mississippi	Jackson
Mo.	MO	Missouri	Jefferson City
Mont.	MT	Montana	Helena

Abbreviation	Zip code abbreviation	State	State capital
Nebr.	NE	Nebraska	Lincoln
Nev.	NV	Nevada	Carson City
N.H.	NH	New Hampshire	Concord
N.J.	NJ	New Jersey	Trenton
N. Mex.	NM	New Mexico	Santa Fé
N.Y.	NY	New York	Albany
N.C.	NC	North Carolina	Raleigh
N. Dak.	ND	North Dakota	Bismarck
–	OH	Ohio	Columbus
Okla.	OK	Oklahoma	Oklahoma City
Oreg.	OR	Oregon	Salem
Pa.	PA	Pennsylvania	Harrisburg
R.I.	RI	Rhode Island	Providence
S.C.	SC	South Carolina	Columbia
S. Dak.	SD	South Dakota	Pierre
Tenn.	TN	Tennessee	Nashville
Tex.	TX	Texas	Austin
–	UT	Utah	Salt Lake City
Vt.	VT	Vermont	Montpelier
Va.	VA	Virginia	Richmond
Wash.	WA	Washington	Olympia
W. Va.	WV	West Virginia	Richmond
Wis.	WI	Wisconsin	Madison
Wyo.	WY	Wyoming	Cheyenne

Provinces and territories of Canada

Abbreviation	State	Capital
Provinces		
AB	Alberta	Edmonton
BC	British Columbia	Victoria
MB	Manitoba	Winnipeg
NB	New Brunswick	Fredericton
NF	Newfoundland	St John's
NS	Nova Scotia	Halifax
ON	Ontario	Toronto
PE	Prince Edward Is.	Charlottetown
PQ	Quebec	Quebec
SK	Saskatchewan	Regina
Territories		
NT	Northwest Territories	Yellowknife
YT	Yukon Territory	Whitehorse

States and territories of Australia

Abbreviation	State	Capital
States		
NSW	New South Wales	Sydney
QLD	Queensland	Brisbane
SA	South Australia	Adelaide
TAS	Tasmania	Hobart
VIC	Victoria	Melbourne
WA	Western Australia	Perth
Territories		
ACT	Australian Capital Territory	Canberra
NT	Northern Territory	Darwin

United Nations organizations

FAO	Food and Agricultural Organization
GATT	General Agreement on Tariffs and Trade
IBRD	International Bank for Reconstruction and Development
ICAO	International Civil Aviation Organization
IDA	International Development Association
IFC	International Finance Corporation
IFAD	International Fund for Agricultural Development
ILO	International Labour Organization
IMO	International Maritime Organization
IMF	International Monetary Fund
ITU	International Telecommunication Union
UN	United Nations
UNCHS	United Nations Centre for Human Settlements
UNICEF	United Nations Children's Emergency Fund
UNCTAD	United Nations Conference on Trade and Development
UNDRO	United Nations Disaster Relief Coordinator
UNEP	United Nations Educational, Scientific and Cultural Organization
UNEP	United Nations Environment Programme
UNFPA	United Nations Fund for Population Activities
UNESCO	United Nations Educational, Scientific and Cultural Organization
UNHCR	United Nations High Commission for Refugees
UNITAR	United Nations Institute for Training and Research

International trade and economic organizations
*(see also **UN organizations**)*

ASPAC	Asian and Pacific Council
ASEAN	Association of Southeast Asian Nations
CARICOM	Caribbean Common Market
CE	*Communauté Eurpéenne*
CERN	*Conseil Européen pour la Recherche Nucléaire*
CSCE	The Helsinki Conference on Security and Cooperation in Europe
EC	European Community
ECOWAS	Economic Community of West African States
EEC	European Economic Community
EFTA	European Free Trade Association
ESA	European Space Agency
IADB	Inter-American Development Bank
ALADI	Latin American Integration Association
NAFTA	North American Free Trade Area
NATO	North Atlantic Treaty Organization
OCAM	*Organisation Commune Africaine et Malgache*

Geographic abbreviations

E	East	**NE**	North east
N	North	**NW**	North west
S	South	**SE**	South east
W	West	**SW**	South west

OECD	Organization for Economic Cooperation and Development
OAU	Organization of African Unity
OAS	Organization of American States
OAPEC	Organization of Arab Petroleum Exporting Countries
OCAS	Organization of Central American States
OIC	Organization of Islamic Conference
OPEC	Organization of the Petroleum Exporting Countries
PTA	Preferential Trade Area for East and Southern Africa
SADC	Southern African Development Community
SPEC	South Pacific Bureau for Economic Cooperation
SPC	South Pacific Commission
SPF	South Pacific Forum
USM	Unlisted Securities Market

Languages

Abb.	Language	Abb.	Language	Abb.	Language
Chin	Chinese	Heb	Hebrew	Port	Portuguese
Eng	English	Hung	Hungarian	Russ	Russian
Fr	French	Ital	Italian	Span	Spanish
Ger	German	Jap	Japanese	Swed	Swedish
Gr	Greek	Lat	Latin		

Books of the Bible

Abb.	Book	Abb.	Book
OT	**Old Testament**	**OT**	**Old Testament**
Gen.	Genesis	**Ps., Pss.**	Psalms
Exod., Ex.	Exodus	**Prov.**	Proverbs
Lev.	Leviticus	**Eccles.**	Ecclesiastes
Num.	Numbers	**Song of Sol.,**	Song of
Deut.	Deuteronomy	**S. of S.**	Solomon
Josh.	Joshua	**Isa.**	Isaiah
Judg., Jud.	Judges	**Jer.**	Jeremiah
Ruth	Ruth	**Lam.**	Lamentations
1 Sam.	1 Samuel	**Ezek.**	Ezekiel
2 Sam.	2 Samuel	**Dan.**	Daniel
1 Kgs.	1 Kings	**Hos.**	Hosea
2 Kgs.	2 Kings	**Joel**	Joel
1 Chron.	1 Chronicles	**Amos**	Amos
2 Chron.	2 Chronicles	**Obad.**	Obadiah
Ezra.	Ezra	**Jon.**	Jonah
Neh.	Nehemiah	**Mic.**	Micah
Esther	Esther	**Nah.**	Nahum
Job	Job	**Hab.**	Habakkuk

Editions of the Bible

AV	Authorized Version	**NIV**	New International Version
GNB	Good News Bible		
NEB	New English Bible	**JB**	Jerusalem Bible
REB	Revised English Bible	**NRSV**	New Revised Standard Version
RSV	Revised Standard Version	**TLB**	The Living Bible
		VULG.	Vulgate

Abb.	Book	Abb.	Book
OT	**Old Testament**	**NT**	**New Testament**
Zeph.	Zephaniah	**Phil.**	Philippians
Hag.	Haggai	**Col.**	Colossians
Zech.	Zechariah	**1 Thess.**	1 Thessalonians
Mal.	Malachi	**2 Thess.**	2 Thessalonians
		1 Tim.	1 Timothy
NT	**New Testament**	**2 Tim.**	2 Timothy
Matt.	Matthew	**Tit.**	Titus
Mk.	Mark	**Philem.**	Philemon
Lk.	Luke	**Heb.**	Hebrews
Jn.	John	**Jas.**	James
Acts	Acts of the	**1 Pet.**	1 Peter
	Apostles	**2 Pet.**	2 Peter
Rom.	Romans	**1 Jn.**	1 John
1 Cor.	1 Corinthians	**2 Jn.**	2 John
2 Cor.	2 Corinthians	**3 Jn.**	3 John
Gal.	Galations	**Jude.**	Jude
Eph.	Ephesians	**Rev.**	Revelation

Scientific abbreviations and constants

Symbol	Name	Unit abbreviation	Unit
a	acceleration	m s^{-2}	metre per second squared
n	amount of substance	mol	mole
L	angular momentum	kg m^2 s^{-1}	kilogram metre squared per second
A	area	m^2	square metre
u	atomic mass unit		
Z	atomic number (proton number)		
C	capacitance	F	farad
ρ	density	kg m^{-3}	kilogram per cubic metre
Q	electric charge	C	coulomb
G	electric conductance	S	siemens
I	electric current	A	ampere
E	electrical field strength (electric force)	V m^{-1}	volts per metre
Ψ	electric flux	C	coulomb
V	electric potential (potential difference: voltage)	V	volt
R	electric resistance		
E	electromotive force	Ω	ohm
W	energy	J	joule
Q	enthalpy	J	joule
S	entropy	J K^{-1}	joule per kelvin
f	focal distance	m	metre
F	force	N	newton
ν, f	frequency	Hz	hertz
g	gravitation field strength	m s^{-2}	metre per second squared
E	heat		
C	heat capacity	J K^{-1}	joule per kelvin
I	kinetic energy	J	joule
L	inductance	H	henry
l	length	m	metre
I	luminous intensity	cd	candela
I	magnetic flux	Wb	weber
B	magnetic flux density	T	tesla

Symbol	Name	Unit abbreviation	Unit
m	mass	kg	kilogram
A	mass number (nucleon number)		
I	moment of inertia	kg m^2	kilogram metre squared
P	momentum	kg m s^{-1}	kilogram metre per second
N	neutron number		
μ	permeability	H m	henry per metre
ϵ	permittivity	F m	farad per metre
V	potential energy		
P	power	W	watt
p	pressure	Pa	pascal
R	reaction	N	newton
n	refractive constant		
A_r	relative atomic mass (atomic weight)		
d	relative density		
M_r	relative molecular mass (molecular weight)		
c	specific heat capacity	J kg^{-1} K^{-1}	joule per kilogram per kelvin
C	thermal capacity	J K^{-1}	joule per kelvin
l	specific latent thermal capacity	J kg^{-1}	joule per kilogram
T	temperature	°C	degree celsius
T	thermodynamic temperature	K	kelvin
t	time	s	second
T	torque	N m	newton metre
W	weight	N	newton
v	velocity	m s^{-1}	metre per second
c	velocity of light	m s^{-1}	metre per second
η	viscosity	Pa s	pascal second
V	volume	m^3	cubic metre
λ	wavelength	m	metre
W	work	J	joule

Units of measurement

Abb.	Units	Abb.	Units
°C	Celsius, centigrade	km	kilometre
cal	calorie	kW	kilowatt
cm	centimetre	l	litre
db	decibel (s)	lb	pound (weight)
°F	Fahrenheit	m	metre
ft	foot	mi	mile
g	gram	mg	milligram
gall	gallon	MHz	megahertz
ha	hectare	min	minimum
hp	horsepower	ml	millilitre
Hz	hertz	mm	millimetre
in	inch	mph	miles per hour
k.	karat	oz	ounce
kcal	kilocalorie	pt	pint
kg	kilogram	V	volt
kHz	kilohertz	yd	yard

Physical constants

Abb.	Constant	Value
1 atm	Standard atmosphere	101 kPa; 760 mmHg
e	Electron charge	1.602×10^{-19} C
c	Velocity of light in a vacuum	2.988×10^{8} ms^{-1}
F	Faraday constant	9.649×10^{4} C mol^{-1}
g	Acceleration due to gravity	9.81 ms^{-2}
G	Gravitational constant	6.667×10^{-11} N m^{-2} kg^{-2}
m_e	Electron rest mass	9.11×10^{-31} kg
h	Planck constant	6.626×10^{-34} J s
N	Avogadro constant	6.023×10^{23} mol^{-1}

Time

Abb.	Unit	Abb.	Unit
h	Hour	s.	Second
min.	Minute		Week
M.	Month		Year
Days			
Mon.	Monday	Fri.	Friday
Tues.	Tuesday	Sat.	Saturday
Wed.	Wednesday	Sun.	Sunday
Thur.	Thursday		
Months			
Jan.	January	Jul.	July
Feb.	February	Aug.	August
Mar.	March	Sept.	September
Apr.	April	Oct.	October
May	May	Nov.	November
Jun.	June	Dec.	December

Speed

Abb.	Imperial	Abb.	Metric
In/s	inch per second	mm/s	millimetres per second
In/min	inch per minute	mm/min	millimetres per minute
In/h	inch per hour	mm/h	millimetres per hour
ft/s	feet per second	cm/s	centimetres per second
ft/min	feet per minute	cm/min	centimetres per minute
ft/h	feet per hour	cm/h	centimetres per hour
mps	miles per second	m/s	metres per second
mpmin	miles per minute	m/min	metres per minute
mph	miles per hour	m/h	metres per hour
		km/s	kilometres per second
		km/min	kilometres per minute
		km/h	kilometres per hour

Medical abbreviations

AA	Alcoholics Anonymous		anaesthesia or evacuation
AC	(*ante cibos*) before food		under anaesthesia
AID	artificial insemination	**FSH**	follicle-stimulating
	by an anonymous donor		hormone
AIDS	acquired immune	**GH**	growth hormone
	deficiency syndrome	**GHRF**	growth hormone releasing
AIH	artificial insemination		factor
	with semen from husband	**GHRIH**	growth hormone release
ARDS	adult respiratory distress		inhibiting hormone
	syndrome	**GnRH**	gonadotrophin-releasing
BMA	British Medical		hormone
	Association	**GOT**	glutamate oxaloacetate
CT scan	Computerized		transaminase
	tomography scanning	**GPI**	general paralysis of the
DNA	deoxyribonucleic acid		insane
DNR	do not resuscitate	**HRT**	hormone replacement
EBV	Epstein-Barr virus		therapy
ECG	electrocardiogram	**IDDM**	insulin dependent
EDTA	ethylene-diamine-		diabetes mellitus
	tetraacetic acid		(Type I)
EEG	electroencephalogram	**Ig**	immunoglobulin
EMG	electromyogram	**IQ**	intelligence quotient
ENT	Ear, Nose and Throat	**IUD**	intrauterine contraceptive
	(Otorhinolaryngology)		device
ERCP	endoscopic retrograde	**IVC**	inferior vena cava
	cholangiopancreatography	**IVP**	intravenous pyelography
ERG	electroretinogram	**IVU**	intravenous urography
ERPC	evacuation of retained	**LDL**	low density lipoprotein
	products of conception	**LMP**	last menstrual period
ESP	extrasensory perception	**LOA**	left occipito-anterior
ESR	erythrocyte sedimentation	**LSD**	lysergic acid
	rate (blood sedimentation		diethylamide
	rate)	**MHC**	major histocompatibility
EUA	examination under		complex

MLD	minimum lethal dose	**RSI**	repetitive strain injury
MRC	Medical Research Council	**SADS**	seasonal affective disorder
MRI	magnetic resonance imaging	**SI units**	units of the *Système International*
MS	multiple sclerosis	**SIADH**	syndrome of inappropriate antidiuretic hormone
NA	Narcotics Anonymous		
NAD	nothing abnormal detected	**SIDS**	sudden infant death syndrome (cot death)
NMR	nuclear magnetic resonance	**SIV**	simian immunodeficiency virus
NSAIDs	non-steroidal anti-inflammatory drugs	**SLE**	systemic lupus erythematosus
OTC	over the counter	**staph**	*staphylococcus, staphylococci,* staphylococcal
PABA	para-aminobenzoic acid		
PEEP	positive end-expiratory pressure	**strep**	*streptococcus, streptococci* or streptococcal
PID	prolapsed intervertebral disc or pelvic inflammatory disease	**TENS**	transcutaneous electrical nerve stimulation
PMS	premenstrual syndrome	**THC**	tetrahydrocannabinol
PMT	premenstrual tension	**TMJ**	temporormandibular joint
Polio	poliomyelitis		
PUVA	psoralens and ultraviolet A	**TPA**	tissue-plasminogen activator
QID	(*quater in die*) four times a day		
QS	(*quantum sufficit*) a sufficient quantity	**TRH**	thyrotrophin releasing hormone
RCM	Royal College of Midwives	**TURP**	transurethral resection of the prostate
RCN	Royal College of Nursing	**VD**	veneral disease
RDA	recommended daily allowance	**VDRL**	Veneral Disease Research Laboratory
RHA	Regional Health Authority	**WBC**	white blood cell
RNA	ribonucleic acid	**WHO**	World Health Organization

Everyday abbreviations

Abb.	Term	Abb.	Term
AC	alternating current; account; air conditioning	CD	compact disc; certificate of deposit
AD	*Anno Domini*, in the Year of Our Lord	ch, chap.	chapter
		Cf.	*confer*; compare
adj.	adjective	CFC	chloroflurocarbon
Adm.	admiral	co.	company; country
adv.	adverb	c/o	care of
advt.	advertisement	coll, colloq.	colloquial
am	*ante meridiem*, before noon	Con.	Conservative
anon.	anonymous	conj.	conjunction
ASAP	as soon as possible	cont. contd.	continued
assn, assoc.	association	Corp.	corporation
		cu	cubic
b.	born	cum	cummulative
B.C	before Christ	CV	*curriculum vitae*
BO	box office	d.	died
BST	British summer time	DC	direct current
C	Roman numeral for 100; century	dec.	deceased
		Dem.	Democrat
c./c.	*circa*/cent; copyright	dept.	department
		ed.	edition; edited; editor
cap.	capital; capital letter	e.g.	*exempli gratia*; for example
CB	citizens' band (radio)	eq.	equal; equation
		esp.	especially

Abb.	Term	Abb.	Term
est.	estimate	HMS	Her (or His) Majesty's Ship
et al.	*et alii*, and others	HQ	headquarters
ETA	estimated time of arrival	ht.	height
ETD	estimated time of departure	I.	island
etc.	*et cetera*, and so forth	ibid.	*ibidem*, in the same place
et. seq.	*et sequens*, and the following	ID	identification
		i.e.	*id est*, that is
FA	Football Association	ill.	illustrated
ff.	following (pages)	inc.	incorporated; including
FM	frequency modulation	Is.	islands
		IQ	intelligence quotient
fig.	figure	ISBN	international standard book number
f/t	full time		
GBH	grievous bodily harm	Jr.	junior
		Lab.	Labour
GMT	Greenwich Mean Time	lat.	latitude
		l.c.	lower case (printing)
govt.	government		
Gen.	general	LCD	lowest common denominator
HGV	heavy goods vehicle		
		Lib.	Liberal
hi-fi	hi fidelity	log.	logarithm
HM	headmaster; headmistress	LP	long player

Everyday abbreviations (continued)

Abb.	Term	Abb.	Term
Ltd.	limited	pa	*per annum*, each year
m.	married	pen.	peninsula
M	medium	per cent.	*per centum*, by the hundred
max.	maximum		
min.	minute; minimum	pdq	pretty damn quick
misc.	miscellaneous	PE	physical education
MO	money order; *modus operandi*, mode of operation	pl.	plural
		plc	public limited company
MS.	manuscript		
Mt	mount; mountain	pm	*post meridiem*, afternoon
n.	noun		
N	north; national; navy	PO	post office
		POB	post office box
N/A	not applicable	pop.	population
N.B.	*nota bene*, note well	pp.	pages
		p.p.	*per procurationem* (by proxy)
nn	notes		
no.	*numero*, number	PS	*post scriptum*, postscript
nr	near		
ob.	*obit*, died	p/t	part time
OK	correct	PTO	please turn over
op.	*opus*, work	QED	*quod erat demonstratum*, which was to be shown
op. cit.	*opere citato*, in the work cited		
o/t	overtime		
p.	page	q.v.	*quod vide*, which see
P	parking		

Abb.	Term	Abb.	Term
R	river	temp.	temperature
R	*Rex*, king; *Regina*, queen	tr.	transpose; translation
R & B	rhythm and blues	UFO	unidentified flying object
rd.	road		
ref.	reference	univ.	university
Rep.	Republican	v.	verb
RIP	*requiescat in pace*, rest in peace	v., vid.	*vide*, see
		v., vs.	versus
Rom.	Romans	VHF	very high frequency
rpm	revolutions per minute		
		VIP	very important person
RSVP	*répondez s'il vous plaît* (Fr.), reply if you please	viz.	*videlicet*, namely
		vol.	volume
S	south; saint; Socialist; senate; society; small	wt	weight
		Xmas	Christmas
		zool.	zoology
s.a.e.	stamped addressed envelope		
sing.	singular		
SLR	single lens reflex camera		
soc.	society		
sp.	species; spelling		
sq.	square		
SS	steamship		
St.	saint; strait; street		